Lecture Notes in Computer Science 8381

Commenced Publication in 1973
Founding and Former Series Editors:
Gerhard Goos, Juris Hartmanis, and Jan van Leeuwe

Aastha Madaan   Shinji Kikuchi
Subhash Bhalla (Eds.)

Databases in Networked Information Systems

9th International Workshop, DNIS 2014
Aizu-Wakamatsu, Japan, March 24-26, 2014
Proceedings

Springer

Volume Editors

Aastha Madaan
Shinji Kikuchi
Subhash Bhalla

University of Aizu
Graduate Department of Computer and Information Systems
Ikki Machi, Aizu-Wakamatsu, Fukushima 965-8580, Japan

Email:
d8131102@u-aizu.ac.jp
shinji.kikuchi@kne.biglobe.ne.jp
bhalla@u-aizu.ac.jp

ISSN 0302-9743   e-ISSN 1611-3349
ISBN 978-3-319-05692-0   e-ISBN 978-3-319-05693-7
DOI 10.1007/978-3-319-05693-7
Springer Cham Heidelberg New York Dordrecht London

Library of Congress Control Number: 2014933394

LNCS Sublibrary: SL 3 – Information Systems and Application, incl. Internet/Web
and HCI

Typesetting: Camera-ready by author, data conversion by Scientific Publishing Services, Chennai, India

Printed on acid-free paper

Springer is part of Springer Science+Business Media (www.springer.com)

Preface

Business data analytics in astronomy and sciences depends on computing infrastructure. Such scientific exploration is beneficial for large-scale public utility services, either directly or indirectly. Many research efforts are being made in diverse areas, such as big data analytics and cloud computing, sensor networks, and high-level user interfaces for information accesses by users. Government agencies in many countries plan to launch facilities in education, health-care, and information support as a part of an e-government initiative. In this context, information interchange management has become an active research field. A number of new opportunities have evolved in design and modeling based on new computing needs of the users. Database systems play a central role in supporting networked information systems for access and storage management aspects.

The 9th International Workshop on Databases in Networked Information Systems (DNIS) was held during March 24–26, 2014, at the University of Aizu in Japan. The workshop program included research contributions and invited contributions. A view of the research activity in information interchange management and related research issues was provided by the sessions on related topics. The keynote address has been contributed by - Prof. Divyakant Agrawal. The section on "Astronomical Data Management" has an invited contribution from Dr. Florin Rusu. The following section on "Business Data Analytics and Visualization," has an invited contribution from Prof. Marcin Paprzycki. The section on "Business Data Analytics in Sciences," includes the invited contributions of Dr. Lukas Pichl. The section on "Business Data Analytics in Astronomy," has an invited contribution by Prof. Thomas A. Prince. I would like to thank the members of the Program Committee for their support and all authors who considered DNIS 2014 for presenting research contributions.

The sponsoring organizations and the Steering Committee deserve praise for the support they provided. A number of individuals contributed to the success of the workshop. I thank Dr. Umeshwar Dayal, Prof. J. Biskup, Prof. D. Agrawal, Dr. Cyrus Shahabi, Prof. T. Nishida, and Prof. Shrinivas Kulkarni for providing continuous support and encouragement.

The workshop received invaluable support from the University of Aizu. In this context, I thank Prof. Shigeaki Tsunoyama, President of the University of Aizu. Many thanks are also extended to the faculty members at the university for their cooperation and support.

March 2014

A. Madaan
S. Kikuchi
S. Bhalla

Organization

DNIS 2014 was organized by the Graduate Department of Information Technology and Project Management, University of Aizu, Aizu-Wakamatsu, Fukushima, PO 965-8580, (JAPAN).

Steering Committee

Divyakant Agrawal	University of California, USA
Hosagrahar V. Jagadish	University of Michigan, USA
Umeshwar Dayal	Hewlett-Packard Laboratories, USA
Masaru Kitsuregawa	University of Tokyo, Japan
Toyoaki Nishida	Kyoto University, Japan
Krithi Ramamritham	Indian Institute of Technology, Bombay, India
Cyrus Shahabi	University of Southern California, USA

Executive Chair

N. Bianchi-Berthouze   University College London, UK

Program Chair

S. Bhalla   University of Aizu, Japan

Publicity Committee Chair

Shinji Kikuchi   University of Aizu, Japan

Publications Committee Chair

Aastha Madaan   University of Aizu, Japan

Program Committee

D. Agrawal	University of California, USA
V. Bhatnagar	University of Delhi, India
P. Bottoni	University La Sapienza of Rome, Italy
L. Capretz	University of Western Untario, Canada
Richard Chbeir	Bourgogne University, France
G. Cong	Nanyang Technological University, Singapore

U. Dayal	Hewlett-Packard Laboratories, USA
Pratul Dublish	Microsoft Research, USA
Fernando Ferri	IRPPS - CNR, Rome, Italy
W.I. Grosky	University of Michigan-Dearborn, USA
J. Herder	University of Applied Sciences, Fachhochschule Düsseldorf, Germany
H.V. Jagadish	University of Michigan, USA
Sushil Jajodia	George Mason University, USA
Q. Jin	Waseda University, Japan
A. Kumar	Pennsylvania State University, USA
A. Mondal	Xerox Research, Bangaloru, India
K. Myszkowski	Max-Planck-Institut für Informatik, Germany
Alexander Pasko	Bournemouth University, UK
L. Pichl	International Christian University, Tokyo, Japan
P.K. Reddy	International Institute of Information Technology, Hyderabad, India
C. Shahabi	University of Southern California, USA
M. Sifer	Sydney University, Australia
F. Wang	Microsoft Research, USA

Sponsoring Institution

Center for Strategy of International Programs, University of Aizu
Aizu-Wakamatsu City, Fukushima P.O. 965-8580, Japan.

Abstracts

Data Exploration in Large Area Time-Domain Sky Surveys: Current Practice and Future Needs

Thomas A. Prince

Jet Propulsion Laboratory
California Institute of Technology
MS 169-327, 4800 Oak Grove Dr., Pasadena, CA 91109
prince@srl.caltech.edu

Abstract. Time-domain astronomical surveys are playing an increasingly important role in astronomy, driven by the availability of large format digital sensors and powerful computational resources. Examples of such surveys include the Palomar Transient Factory (PTF), the Catalina Real-time Transient Survey (CRTS), the PanSTARRS survey, and the future Large Synoptic Survey Telescope (LSST) and Zwicky Transient Facility (ZTF). These surveys typically are sensitive to on order of a billion or more objects over the course of a year and the challenge is to find the few hundred to a few thousand most interesting variable sources among the total collection of objects. These include supernovae, gamma-ray bursts, near-earth asteroids, cataclysmic variables, and a large number of other exotic objects. While technical challenges exist in the area of database management and storage, the most difficult challenges are algorithmic. We will discuss some of the principal challenges, including the unique problems associated with the intrinsically heterogeneous collection of time-domain data and the application of machine learning techniques to automatic identification of interesting variable sources.

Towards an Intelligent Astronomical Event Broker: Automated Transient Classification and Follow-Up Optimization

Przemek Wozniak, Ph.D.,
Optical Sensing Team Leader

Space and Remote Sensing, ISR-2, MS-B244
Los Alamos National Laboratory, Los Alamos, NM 87545
wozniak@lanl.gov

Abstract. In order to succeed, the massive time-domain surveys of the future must automatically identify actionable information from the torrent of imaging data, classify emerging events, and optimize the follow-up strategy. The impedance mismatch between the high rate of transient detection and limited follow-up resources is growing rapidly. Despite a pressing need to delegate increasingly complex data fusion and inference tasks to machines, the available technology toolbox remains underutilized in astronomy. To address these challenges, we are developing an autonomous, distributed event broker that will integrate cutting edge machine learning algorithms with high performance computing infrastructure. The talk will give an overview of various efforts in this area, including recent progress on image level variability detection, spectral classification using low resolution spectra and dynamic coalition management approach to follow-up.

Machine-Learning Enabled Stellar Classification and the Prediction of Fundamental Atmospheric Parameters From Photometric Light Curves

Adam A. Miller

Jet Propulsion Laboratory
California Institute of Technology
MS 169-327, 4800 Oak Grove Dr., Pasadena, CA 91109
amiller@astro.caltech.edu

Abstract. The falling costs of computing and CCD detectors has led to a great boom in wide-field time-domain surveys during the past decade, with several new surveys expected prior to the arrival of the Large Synoptic Survey Telescope (LSST). This observational boon, however, comes with a catch: the data rates from these surveys are so large that discovery techniques heavily dependent on human intervention are becoming unviable. In this talk I will detail new methods, which utilize semi-supervised machine-learning algorithms, to automatically classify the light curves of time-variable sources. Using these methods, we have produced a data-driven probabilistic catalog of variables found in the All Sky Automated Survey (ASAS). I will also present a new machine-learning-based framework for the prediction of the fundamental stellar parameters, Teff, log g, and [Fe/H], based on the photometric light curves of variable stellar sources. The method was developed following a systematic spectroscopic survey of stellar variability. I will demonstrate that, for variable sources, the machine-learning model can determine Teff, log g, and [Fe/H] with a typical scatter of 130 K, 0.38 dex, and 0.26 dex, respectively, without obtaining a spectrum. Instead, the random-forest-regression model uses SDSS color information and light-curve features to infer stellar properties. The precision of this method is competitive with what can be achieved with low-resolution spectra. These results are an important step on the path to the efficient and optimal extraction of information from future time-domain experiments, such as LSST. We argue that this machine-learning framework, for which we outline future possible improvements, will enable the construction of the most detailed maps of the Milky Way ever created.

Astrophysical Image Modeling

Robert M. Quimby

Kavli Institute for the Physics and Mathematics of the Universe (WPI)
Todai Institutes for Advanced Study
the University of Tokyo
5-1-5 Kashiwa-no-Ha, Kashiwa City, Chiba 277-8583, Japan
robert.quimby@ipmu.jp

Abstract. Discovery time variable objects in astronomical data requires that the same patch of sky be observed at least twice and preferably more often. The first observation can be used to predict what that same patch of sky will look like in subsequent observations to finite accuracy (given noise and changes in the atmosphere or instrument through which the observations are made). In other words, a model can be constructed by some transformation of the data obtained at the first epoch to match the observational characteristics of a second observation. Comparing this model to the actual observations, candidates for time variability will stand out as objects that are poorly predicted by the model. Here I discuss how these models of the astrophysical sky are currently constructed using the Palomar Transient Factory as a specific example. I will point out some of the weak points of this process and discuss some new techniques which may prove beneficial for the discovery of time variable objects.

Table of Contents

Big Data and Cloud Computing

Astronomical Data Management

Business Data Analytics and Visualization

Business Data Analytics in Sciences

Geo-spatial Decision Making and Query Languages

Big Data in Online Social Networks: User Interaction Analysis to Model User Behavior in Social Networks

Divyakant Agrawal, Ceren Budak, Amr El Abbadi,
Theodore Georgiou, and Xifeng Yan

Department of Computer Science,
University of California, Santa Barbara
{agrawal,cbudak,amr,teogeorgiou,xyan}@cs.ucsb.edu

Abstract. With hundreds of millions of users worldwide, social networks provide incredible opportunities for social connection, learning, political and social change, and individual entertainment and enhancement in a multiple contexts. Because many social interactions currently take place in online networks, social scientists have access to unprecedented amounts of information about social interaction. Prior to the advent of such online networks, these investigations required resource-intensive activities such as random trials, surveys, and manual data collection to gather even small data sets. Now, massive amounts of information about social networks and social interactions are recorded. This wealth of big data can allow social scientists to study social interactions on a scale and at a level of detail that has never before been possible. Our goal is to evaluate the value of big data in various social applications and build a framework that models the cost/utility of data. By considering important problems such as Trend Analysis, Opinion Change and User Behavior Analysis during major events in online social networks, we demonstrate the significance of this problem. Furthermore, in each case we present scalable techniques and algorithms that can be used in an online manner. Finally, we propose the big data value evaluation framework that weighs in the cost as well as the value of data to determine capacity modeling in the context of data acquisition.

Keywords: Social Networks, Big Data, Social Analytics, Data Streams, Complex Networks.

1 Introduction

One of the main challenges confronting researchers in many diverse fields is the analysis and understanding of very large data sets. Not only do physical scientists face this challenge when observing natural phenomena or studying experimental results, but social scientists are also being exposed to ever increasing and diverse data sets. In spite of the challenges associated with big data, this phenomenon is enabling scientists approach traditional problems from new perspectives. In the

A. Madaan, S. Kikuchi, and S. Bhalla (Eds.): DNIS 2014, LNCS 8381, pp. 1–16, 2014.
© Springer International Publishing Switzerland 2014

context of social sciences prior to the advent of online networks, various investigations required resource-intensive activities such as random trials, surveys, and manual data collection to generate even small data sets. Now, many social interactions take place in an online environment, and as a result, massive amounts of data about social networks and social interactions are recorded. This wealth of data, presenting an almost-natural yet not easily controllable laboratory for social experiments, can allow social scientists to study social interactions at a scale and at a level of detail that has never been possible before. In fact, it has been argued that online social networks present social scientists with a unique opportunity to observe and analyze interactions in social networks. The right set of *data summarization* tools can help scientists extract *knowledge* out of these ever increasing and diverse data sets.

Modern on-line social networks, such as Facebook, Twitter, and Renren contain a wealth of public information regarding the interactions, likes and dislikes, interests of hundreds of millions of individuals who form large segments of the global society. Facebook and Twitter each claim about 800 million users, and at any given moment millions of interactions occur among the users of each of these social networks. Communication exchanges are occurring on a continuous basis, with about 500 million tweets per day on Twitter and a peak of 143199 tweets per second observed during the airing of a movie in Japan on August 2013 [1]. Both the topic as well as the pattern of such communications can provide deep insights in diverse and sometimes critical contexts. For example, it was reported that during the 2008 Santa Barbara fires, on-line social networks were considered more reliable and up to date with fire locations and evacuation information than the traditional media outlets; or that tweets often spread faster than the tremors of an earthquake [46]. The benefits of online social networks during emergency events extends to natural disasters such as hurricanes and earthquakes [21,51]. In general, in emergency situations both the content as well as the spread of information in social networks can provide valuable *knowledge* that is critical in life saving situations.

The utility of online social networks is not limited to emergency events. Recent evidence indicates that 45% of users in the U.S. say that the Internet played a crucial or important role in at least one major decision in their lives in the last two years, such as attaining additional career training, helping themselves or someone else with a major illness or medical condition, or making a major investment or financial decision [19]. In fact, basic human activities have changed in the context of the Internet and social networks, and new possibilities have emerged. For instance, the process by which people locate, organize, and coordinate groups of individuals with shared interests, the number and nature of information and news sources available, and the ability to solicit and share opinions and ideas across myriad topics have all undergone dramatic change as a result of interconnected digital media. Furthermore, increasing reliance on the "wisdom of crowds" has been demonstrated to both solve and effectively predicate diverse human behavior. Aggregating the efforts of anonymous crowds has been demonstrated to help address complex issues [20,36]. There is also growing evidence that communities and the exchange of information among connected individuals result in increasing the overall knowledge of the community. Capturing this knowledge is a major challenge [31], and if accurately captured, channeled and articulated, it can help solve many of humanities challenges, such as harnessing human capacity to overcome such endemic challenges as world hunger and illiteracy. This wealth of information, though present in social networks, is buried under a large amount of noise in big data. Even in the cases where the vastness of data is not necessarily a disadvantage, the advantage, or rather the amount of it, needs to be questioned. Lately, there has been growing rhetoric that argues that the information you can extract from any big data asymptotically diminishes as your data volume increases [58].

So, is more data always better? As counter intuitive as it sounds, the answer to this question is not simply "yes". Instead, the answer, while being less satisfactory, is "depends". For instance, the value of *big* data in identifying correlations between two measures x and y in a data set is questionable [14]. It's not hard, even with a data set that includes just 1,000 items, to get into a situation in which we are dealing with many, many millions of correlations. This means that out of all these correlations, a few will be extremely high just by chance: if you use such a correlation for predictive modeling, you will lose [14]. We explore different Social Behavior problems through an analysis of large datasets. We study the problem of Trend Analysis (trending topics) in various levels; from simple trend detection to multi-dimensional trend analysis. We analyze how Opinion changes in a social context and how sentiment varies as global and local opinions change. Finally we explore the area of Event Detection and Summarization and how users behave and break news during large scale and real life events. What is common between all these social behavior problems is that Big Data plays a critical and not always beneficial role. Our long term goal is to further study the impact and implications of Big Data and introduce a framework that can analyze a social problem and attempt to answer the following critical questions: (1) Is a dataset appropriate (utility of the data) and (2) do we need more or less data (amount of data)?

2 Analytic Approaches for User Behavior Modeling

In this section we present related and established work on two of the social behavior applications we study, Trend Analysis and Opinion Change.

2.1 Trend Analysis

Online social networks contain a large variety of information and identifying specific information items of importance has been of interest since their inception. A simple metric for identifying the importance of specific informational topics can be evaluated by the accumulated interest such topics receive from users over time. Such measures are already in use as in the case of YouTube (view counts of videos) or Digg.com (number of *diggs* a story receives). Considering the usefulness and possible impact of this measure, we first start with a formal *count-based trends* in which topics accumulate value over time according to the number of times they have been mentioned. Assume users of a network can choose to (or not to) broadcast their opinions about various topics at any point in time. Assume further that we can abstract away what the topic is from what a user broadcasts. In this setting, we model a *mention* by node n_i on a specific topic T_x as a tuple $\langle n_i, T_x \rangle$. We refer to the history of such tuples as *stream* and denote it using S. Under this model, *count-based* trendiness of T_x can be defined as:

$$f(T_x) = \sum_{n_i \in N} C_{i,x} \tag{1}$$

where $C_{i,x}$ represents the number of mentions of the form $\langle n_i, T_x \rangle$ in S, i.e. the number of mentions by node n_i of topic T_x and N is the set of users in the social network.

The top-k topic detection problem, when the score is defined in this way, is simply to find the frequent items in a stream of data, also referred to as *heavy hitters*. This problem can be easily solved by keeping track of the accumulated count for each topic discussed in the social network. However, for large and dynamic data sets, it is desirable to look for approximate solutions. Given the large scale of online social networks today, both in terms of number of users and volume of activity, even the simple count-based trends detection calls for such approximate solutions. The *frequent elements problem* has been well studied and several scalable, online solutions have been proposed [8,13,43,40]. The algorithms for answering frequent elements queries are broadly divided into two categories: *sketch-based* and *counter-based*. In the sketch-based techniques [8,13], the entire data stream is represented as a summary "sketch" which is updated as the elements are processed. On the other hand, counter-based techniques [43,40] monitor a subset of the stream elements and maintain an approximate frequency count. The Space Saving algorithm, a counter-based algorithm [43], has been identified to have the best throughput amongst its class of frequency counting algorithms [12]. We therefore plan to use it as a building block for discovering count-based trends.

In recent years, there has been a great increase in research relating to on-line social networks. While early works focused on static social networks analysis [32,33], more recent research evolved to study more complex and dynamical notions such as information diffusion [27]. As importance of information trends in social networks increased, there has been a number of studies that focused on information trends from various perspectives [4]. For instance, Kwak et al. [33] study and compare trending topics in Twitter reported by Twitter [56] with those in other media. The results show that the majority of topics are headline news or persistent news in nature. In [35] Leskovec et al. study temporal properties of information shared in social networks by tracking "memes" across the blogosphere.

Recently, a number of works have studied structural properties of graphs in a streaming or semi-streaming fashion. The computation of network indices based on counting the number of certain small subgraphs is a basic tool in the analysis of the structure of large networks. A type of problem that is significantly related to the problem studied here is counting triangles in a graph stream. There are three types of solutions to this problem: exact counting [5], streaming [6] and semi-streaming algorithms [7]. Detecting trends that are not oblivious to the underlying structure of an online social network requires online solutions and therefore these techniques are not directly applicable.

Another important characteristic of news or discussions in social networks is the spatial properties of the agents that are involved in the discussion or the source of the news. A recent work by Teitler et al. [53] collects, analyzes, and displays news stories on a map interface, thus leveraging their implicit geographic context. A follow-up study performs similar techniques to identify geographical information in news in Twitter. Although these works that focus on temporal and spatial characteristics of trends are important for a better understanding of the notion of trends, they are orthogonal to the approaches introduced in this study, as they focus on identifying tweet clusters based on locations and not trend detection. Recently, there has been more effort in online analysis of geo-trends in social networks [38,18]. Hong et al. [18] focus on user profiling from a geographical perspective by modeling topical interests through geo-tagged messages in Twitter. This problem is orthogonal to the problem studied in here as it focuses on user-centric modeling in an offline manner while our approach aims at detecting trends in an online fashion. Similar to our work, MacEachren et al. [38] study the problem of identifying significant events in different localities for crisis management. However, this work provides a high level framework while we provide efficient algorithmic tools with accuracy guarantees.

2.2   Opinion Change

The proliferation of social media, forums, and networks has witnessed the power of networks that propagate news, opinions, and stances on a scale and speed that have never been seen before. Unfortunately, due to the lack of appropriate metrics and models, we are not able to characterize, quantify, and predict persuasion that is occurring everywhere in the social-cyber space. At the core of analyzing persuasion over networks, there is a fundamental problem: how to measure, model and simulate the opinions and shifts of opinion of users in a network, with reasonable accuracy. While it is difficult to derive an accurate model on the individual level, we have demonstrated in our work this year that it is possible to build a collective model over groups of people that share similar opinions over a set of specific topics. By studying and comparing the position and the dynamics of position, persuasion patterns and knowledge that are hidden in complex social and information networks may be revealed.

In order to detect reasons for public opinion change/persuasion, one can first track sentiment variation towards the interested target. If a significant change in crowd sentiment is observed on Twitter, one can analyze tweets during the corresponding period to discover the reasons. There are three challenges for this task: (1) Tweets are very noisy and cover many general topics/events which do not really contribute to sentiment change. How to filter out these unrelated topics/events is a serious issue. Text summarization techniques are not appropriate for this task since text summarization aims at covering all topics/events in the text collection. Similarly, extracting the most frequently mentioned words during the change to represent the reasons is not a good idea, as these words may actually come from the background or general topics/events which have been discussed for a long time. (2) Events sometimes are complex and are composed of a number of small events. The change of opinion may be caused by only one subevent but not the whole event. How to find these fine-grained reasons is generally very challenging. (3) The third challenge is how to properly represent the reasons. Keywords or topics output by Topic Modeling methods [55] can describe the underlying topic to some extent; but this is not as intuitive as natural language sentences.

Topic-based User Sentiment Analysis and Classification. In order to study and analyze the change in users topic sentiments across time, we first must discover their sentiment from communication data. There have been numerous prior sentiment classification methods introduced which focus upon the determination of sentiment (classifying as positive or negative) within messages sent between users on Twitter [41,57,23,16,9] and other social media websites [45,42,50,39,30,49]. We will later describe our existing approach that classifies tweets from a large, real-world Twitter dataset combining Tan et al.'s technique [57], and Mudhakar et al.'s Multinomial Bayes classifier [45].

Modeling of Sentiment Change in Social Networks. Most current models for the spread of ideas and influence in social networks are based on diffusion of the idea from a node to its (directed) neighbors. For example, if the status x(t) at time t of node i is either active (an adopter of the idea) or inactive, then such a model might postulate that the status at the next time interval is given by $x_i(t+1) = \sum w_{ij}x_j(t)$, where the weighted sum is taken over node i and its immediate neighbors. This type of model, which has its roots in social network theory, was developed to explain small group dynamics [15], and has also been used to simulate dynamics of fish schools [25]. Variants of this basic diffusion model are the Voter Model, and Independent Cascade and Linear Threshold Models [26,29,10]. These types of models have been very successful in explaining information propagation, and they lend themselves well to theoretical analysis. On the other hand, they have known shortcomings. For example, according to social network theory, a given node will keep updating its status even if the status of its neighbors is unchanging. This seems unrealistic, particularly in the context of social networks. Our plan is to determine the extent to which diffusion models can be validated on real social network opinion data, and then to consider potential improvements and extensions to the models.

Controlling Opinion and Detecting & Countering Control. A large number of works relating to influence maximization and opinion control on social networks have been previously introduced. Many of these have taken a threshold approach to modeling decision and opinion change of users, theorizing that the current number of neighbors with an opinion decides the current opinion of a node [26,17,47]. This type of threshold approach is therefore timing and order-independent, as the timing of neighboring users' opinion changes does not affect the resulting opinion decisions. It is only recently that works have begun to take into account the effect of ordering within influence and opinion cascades, acknowledging that timing and sequence play a vital role in the spread of opinions. In [11], Chierichetti et al. looked at the effect of sequence among neighbors opinion changes, (determining optimal orderings of sentiment changes) to analyze this effect on product adoption cascades. However, while the sequence of neighboring opinions are studied in [11], the actual timing and rates of these opinion adoptions how closely clustered they are in time are ignored.

3 New Approaches for User Behavior Analysis and Modeling

In this section we describe future research on User Behavior in Social media and the impact that Big Data has in each case.

3.1 Semantic-Based Information Trends

Information that is shared in a social network may have certain semantic properties such as the location and time. For instance, one might be interested to know the trends in California alone or short/long term trends . Such queries cannot be answered using trends analysis at the scale of the entire network. Therefore we believe there is a need for trend definitions that explore such dimensions. Our belief is also supported by the growing body of research in this field [54,52]. In this section, we first discuss trends that explore the spatial and temporal characteristics of data.

Spatial trends can be defined in various ways. For instance, the goal can simply be to detect heavy hitters for each location. However, such a technique fails at identifying topics of true geographical nature since a topic of global importance incidentally also has a high frequency of occurrence in various localities without really being related to such locations. Distinguishing such a topic from ones that are trending in only certain localities is not possible without considering the *correlations* between places and topics. Therefore, we plan to focus on the problem of identifying the correlation of information items with different geographical places. We propose *GeoWatch*: an algorithmic tool for detecting geo-trends in online social networks by reporting *trending* and *correlated* location-topic pairs. *GeoWatch* also captures the temporality of trends by detecting geo-trends along a sliding window. With the use of different window sizes, trends of different time granularity can be detected. Our analysis on a Twitter data set shows that such geo-trend detection can be very important in detecting significant events ranging from emergency situations such as earthquakes to locally popular flash

crowd events such as political demonstrations or simply local events such as con-
certs or sports events. The fast detection of emergency events such as the March
11 Japan earthquake indicates the possible value of *GeoWatch* in crisis management. In Figure 1, we present a heat map of tweets for a period of approximately
2 months of tweets (March 9 to May 8, 2011). More particularly, we capture the
volume of tweets originating from various cities in Figure 1(a) and tweets about
cities in Figure 1(b). In these plots, every city associated with more than 10
tweets is marked– color and size is proportional to the number of tweets. Our
approach helps identify various characteristics of the social network usage. The
two figures resemble each other but there are certain interesting distinctions. It
is worthwhile to note that the part of the map corresponding to Japan is denser
in Figure 1(b). This is mostly due to the Japan Earthquakes that took place
within the time period captured in our data set. This important event spanned
a long time period due to the after effects and was an important headliner, making it a trending topic in Twitter. On the contrary, a drop in significance can be
observed for countries such as Indonesia when comparing the tweets *in* cities to
tweets *about* cities. This big difference originates from the fact that Indonesia
is a highly active country in Twitter [22], while there are no important events
taking place in its cities that would result in people mentioning them. As part of
our proposed work, we aim to provide further insights into the data by making
it clear which localities (in terms of geography and time) are similar in behavior
or which localities play a critical role in a given topic trending. This way, we
not only help users focus on a given localities (or time period) and observe the
trends there, but we also allow users to focus on a given set of topics and look
at them from the perspective of geo-spatial characteristics.

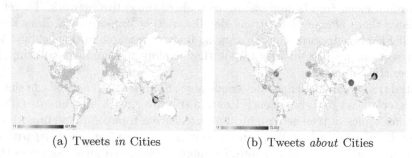

(a) Tweets *in* Cities   (b) Tweets *about* Cities

Fig. 1. Heat Map for # of tweets in/about cities of the world

Problem Definition. Given a stream S of location-topic pairs of the form (l_i, t_j),
a window size of N, and three user defined frequency thresholds θ, ϕ, and ψ in
the interval $[0, 1]$; our goal is to keep track of all locations l_i s.t. $F(l_i) > \lceil \theta N \rceil$
alongside their frequencies as well as all topics t_x and their frequencies $F(t_x)$. In
addition, in order to detect the correlations, we aim to find all pairs (l_i, t_x) s.t.
$F(l_i) > \lceil \theta N \rceil$, $F(l_i, t_x) > \lceil \phi F(l_i) \rceil$, and $F(l_i, t_x) > \lceil \psi F(t_x) \rceil$; where $F(l_i, t_x)$ is
the number of information items on topic t_x from location l_i in the most recent
N items in S; $F(l_i)$ is the aggregate number of occurrences of all the items from

l_i in the current time window; and $F(t_x)$ is the aggregate number of items on t_x. The window size can be set in terms of maximum number of elements or an actual time window such as an hour or a day. In the latter case, the number of elements N in the current window is variable.

Methodology and Data Structures. We now explore a sketch-based structure for *GeoWatch* to detect correlations between locations and topics. The problem of detecting correlations in multi-dimensional datastream has been studied to detect advertising fraud in clickstream data [44]. However, the solution is counting-based and hence only supports insert operations and cannot deal with information deletion. As can be seen from Figure 2, *GeoWatch* consists of two main components. *Location-StreamSummary-Table* contains a $StreamSummary_{l_i}$ structure for each location l_i that has a current estimated relative-frequency of at least θ. In order to provide a solution in a sliding window where deletions as well as insertions of elements need to be supported, *Location-StreamSummary-Table* also needs to include a sketch structure. This sketch structure is maintained to keep track of frequencies of locations in a sliding window by allowing both insertion and deletion operations [24]. In general *GeoWatch* uses sketches to keep track of the frequencies of tracked elements. The second component is the *Topic-StreamSummary-Table*, a hash table that monitors the topics that are potentially correlated with at least one location and a sketch structure to keep track of the topic frequencies. For each tracked topic this structure also keeps track of the number of locations the topic is trendy for. Once this value reaches 0, the topic is removed from *Topic-StreamSummary-Table*.

Even though the development of *GeoWatch* also captures the notion of temporality through the use of sliding windows, its main focus is the spatial characteristics of data. As part of proposed work, we will investigate analyzing information trends at different temporal granularities such as by the minute, hour, days, and so on and doing so in an efficient manner. Furthermore, we aim to identify topics that suddenly become popular, i.e., a topic that is not necessarily a heavy-hitter in the traditional sense but exhibits a sharp increase in frequency over a short period of time. In order to discover such trends, it is necessary to consider both the frequency and the temporal order of elements in a data stream. While many of the data stream algorithms ignore temporal order, there have been several works that have incorporated some notion of the temporal aspect [28,3,2,34].

Big Data Implications. The whole approach we want to take is purely a way to deal with the Big Data nature of the problem. While an exact solution would be 100% accurate, counting and storing all the pairs makes it impossible. Dealing with an information stream that produces thousands of updates per second and being able to report trending pairs in real time dictates approximate counting, space efficient data structures and sliding windows. Now a good question would be if we can sample the data in the stream and get equally good results. Our initial experiments for the proposed approach show that the quality should be nearly perfect but further studying how reducing the data volume can affect the quality or even the proposed algorithms is a very important direction.

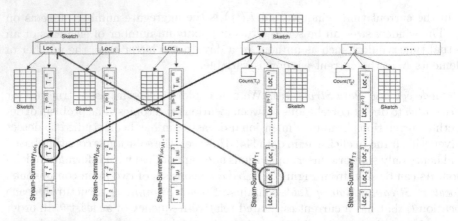

Fig. 2. Overview of Data Structure: The two main sub-components are *Location-StreamSummary-Table* (on the left) and *Topic-StreamSummary-Table* (on the right). *Location-StreamSummary-Table* keeps track of ϕ-frequent topics for each of the θ-frequent locations. *Topic-StreamSummary-Table* keeps track of ψ-frequent locations for each topic that is ϕ-frequent for at least one location. Here the third most important topic for Loc_1 is T_2 and the second most important location for T_2 is Loc_1.

3.2   Multi-dimensional Trend Analysis

A natural extension of the spatio-temporal trend analysis would be to extract trends that focus on multiple dimensions; location and topic being just two of them. There are no limitations on the nature of the dimensions: it can be demographics like age or gender, it can be a location hierarchy, it can be a user's characteristic like opinion, political support or product preference. By analyzing data in a highly dimensional space we can discover trends like "an unusual number of people in the age interval of 18-25, that owns an iPhone, and live in Louisiana mention the topic #CES2013". This information can be extremely valuable to companies, advertisers, political parties and others that need to understand their audience and how they behave, how they are distributed on the map, what topics they are interested in and many other aspects depending on the monitored dimensions. While this problem is the generalization of the spatio-temporal trend analysis described in the previous section, the introduced challenges are not straight forward to solve. Even the exact solution of counting all observed tuples can be very expensive in both computational time and space. And it gets trickier with an online solution where both time and space have to be at most sub-linear.

Big Data Implications.   In this particular problem, having Big Data is both beneficial and problematic. On one hand, the curse of having many dimensions suggests that having more data will result in more dense trends but on the other hand, efficiently counting "interesting" frequent tuples (and not all of them as we do in the Database field of frequent item counting), in an online manner, is very challenging.

3.3   Opinion Change

Opinion change consists a two-fold problem: First, the actual opinion or sentiment has to be identified and then, a change must be observed. Opinions change
around us all the time and studying the behavior of users and how they make
up (or not) their minds can be very useful. As a huge amount of people use social media and express themselves freely, the mining of opinions and how these
change should be a rather easy task but depending on the definition of Opinion
it can be quite the opposite.

While the general definition of Opinion describes it as a viewpoint or statement about a subjective matter, in many research problems we assume more
specific and simpler definitions. For example, sentiment analysis is considered to
be a type of opinion mining even if it's only focused on extracting the sentimental
score from a given text. So assuming a more simplistic definition of Opinion, we
can view people's preferences of political parties or products as opinions. Therefore, there is a wealth of signals to mine from social data like posts on Twitter
of Facebook and extract opinions, identify if they change while time passes and
analyze what events or other factors contribute in these changes. However, different types of opinions require different types of analysis. For example, Twitter
users express their musical preference much more frequently and easily than they
do with politics. Also, the reasons why someone might change their opinion on
Apple products can be very different and less deep than why they would change
they political lean.

Preference in Mobile Devices. The first experiment we conducted while studying
the correlation between opinion change and the contribution the social network
has on it, focused on Twitter users and their preference on mobile devices. Utilizing the tweet's "source" field that indicates the software client where a tweet
was sent from, we were able to build temporal profiles for every user that had a
non trivial amount of tweets. These profiles contain the type of the mobile device
the user is using (iPhone, iPad, Blackberry, Android, Windows Phone) for every
day they tweeted. We assumed this feature as the opinion of a user at any given
time, for their mobile preference. Note that the dataset was quite accurate since
it reflected the actual device a person was using. Using these timeseries we were
able to tell when people switched to a different type of device (e.g. from Android
to iPhone).

Having identified the "change" of opinion we then studied if there was a
network effect in the process of the decision making. We know when people
switched devices and we can also find the account they followed at that time.
By joining the social graph with the opinion signal we were able to compute
the distribution of mobile devices for the neighbors of every user that changed
device. While the hypothesis and generally the literature suggests that people
in one's network have an impact on that person, it was not validated in our
experiment. There wasn't any statistical significance in the observed results and
the fact that we had the complete dataset didn't make a difference. So what we
learned from this experience, is that having a lot of data, no matter how Big

or complicated, can't always make up for information that simply is not there. If people trust more what their real life friends say about phones, then just observing their follow graph can be more misleading than beneficial. There has been some recent work [37] on how it is better to focus on specific groups of users when studying behavior rather than the whole population which is very noisy and can also be quite biased. This further underlines the fact that truncating a dataset in a smart and correct way and reducing Big Data to just Data can be sometimes mandatory for specific social applications. We would like to further explore this direction in a framework that can automatically identify such cases.

3.4   User Behavior during Events

The last application we want to study in this context has to do with real life and real time Events. It has become a second nature these days to talk in social media about things that happen in real life. When something happens there is an information rally from users that try to break the news first, write updates and consume content. Common people at the right place and the right moment can give away information on something that happens, before any news agencies. This gives the chance to literally anyone to have their 10 minutes of fame and also highlights the importance of non power-users in social networks. With the recent events at Boston's Marathon on April 15, 2013 (Boston Marathon bombings), we observed a unique situation where people were live reporting from the scene of crime. In the context of analyzing what is happening and shaping information as in a news feed, it is extremely important to be able to capture such cases as soon as possible. Building an application that can report breaking news requires minimizing the reporting latency while maximizing the recall and accuracy of the reported content. It is easy to wait for a story to appear on major news channels but this compromises the latency in about 50% of the times [48]. On the other hand, extracting breaking news from users that are not priorly known to generate such content may lead to unpredictable and questionable quality.

We are proposing the study of a method that can discover in real time, and as soon as possible, the unique users that for a short time span have a very large reporting value (could be even larger than from a news reporting site). This problem belongs in the area of information diffusion and we can view these people as one-time only innovators where their discovery is a time sensitive task. Being able to assign a breaking score to users based on how other users are consuming their content is expected to be an approach to the right direction. One could view this problem as a trend detection problem where instead of topics, as discussed in the previous sections, we have users. A trending user is a user that trends in terms of consumption of their produced content; we count how many times other people are sharing or just consume what they say (e.g. tweets) in a time-window. A time window approach sounds reasonable since as with most trend analysis applications, trendiness is temporal. Users that are interesting to follow during a specific event, are not so likely to generate interesting content for other types of events, therefore we want their trendiness to decade.

Big Data Implications. While we believe that this problem can be viewed as a trend detection problem, it is unclear if it shares the same properties of the other trend analysis tasks we described. Would the same datasets behave equally good? Do we need extra features to get better quality? We believe that this question is very important and worth study.

4   Research Vision: A Scoring Framework for Big Data

As we discussed in the previous sections, Big Data in the context of social behavior analysis can result to largely different benefits or challenges depending on the nature of the studied problem. While each problem is important on its own, all share the same denominator, the fact that Big Data is used and that it's not clear how we should use it. We are proposing the development of a framework that can score a dataset when given a research problem. While we wouldn't be able to score an unknown dataset for a new problem we can score new datasets on well studied problems (like trend analysis) or give a confidence score on how well an existing dataset may work for a new problem. In both cases we need to identify the features and characteristics that make a dataset suitable for a specific application and then we should be able to extrapolate. We also need to take into consideration the actual cost of obtaining a data set. Data can be extremely large while some problem might require a very specific subsets. Or in other cases, equally good quality can be achieved after sampling. Therefore, a cost model must capture the following salient points:

- Cost of acquisition: Can have different values associated with each column (phone number can be more costly then country of residence, user profile can be more costly then a tweet)
- Cost of storage: Traditional methods could suffice. Can we bring cloud into the picture here?
- Cost of processing: Same here, can we bring the cloud?
- SLA: The nature of social problems usually dictate a fast response. Processing large amounts of data can result in long processing time which might be unacceptable.
- Value: While the first 4 parameters can simply be input to the cost model, the value is harder for a user to pinpoint. Especially working in a probabilistic and highly unpredictable space such as computational social science, it is hard to pinpoint the real value of a solution or data. We propose a data-centric approach to identify the value function in a per-application basis. In particular, we learn from a training data the relationship between value and characteristics such as amount, time and location of data.
- Meta-data on data: We believe that the value of the data is not simply embedded in its amount. The behavior of data changes through time and space.

Acknowledgments. This work is partially supported by NSF Grant IIS-1135389 and a gift from the Bill and Melinda Gates Foundation.

References

1. New tweets per second record, and how!
   https://blog.twitter.com/2013/new-tweets-per-second-record-and-how
2. Aggarwal, C.C., Han, J., Wang, J., Yu, P.S.: A framework for clustering evolving data streams. In: Proc. 29th Int. Conf. on Very Large Data Bases, pp. 81–92. VLDB Endowment (2003)
3. Aggarwal, C.C., Yu, P.S.: Online analysis of community evolution in data streams. In: Proc. SIAM International Data Mining Conference (2005)
4. Allan, J. (ed.): Topic detection and tracking: event-based information organization. Kluwer Academic Publishers, Norwell (2002)
5. Alon, N., Yuster, R., Zwick, U.: Finding and counting given length cycles. Algorithmica 1717, 209–223 (1997)
6. Bar-Yossef, Z., Kumar, R., Sivakumar, D.: Reductions in streaming algorithms, with an application to counting triangles in graphs. In: SODA 2002, pp. 623–632 (2002)
7. Becchetti, L., Boldi, P., Castillo, C., Gionis, A.: Efficient semi-streaming algorithms for local triangle counting in massive graphs. In: KDD 2008, pp. 16–24 (2008)
8. Charikar, M., Chen, K., Farach-Colton, M.: Finding frequent items in data streams. In: Widmayer, P., Triguero, F., Morales, R., Hennessy, M., Eidenbenz, S., Conejo, R. (eds.) ICALP 2002. LNCS, vol. 2380, pp. 693–703. Springer, Heidelberg (2002)
9. Chen, B.: Topic oriented evolution and sentiment analysis. Ph.D. Dissertation, Penn State University (2011)
10. Chen, W., Wang, Y., Yang, S.: Efficient influence maximization in social networks. In: KDD 2009, pp. 199–208 (2009)
11. Chierichetti, F., Kleinberg, J., Panconesi, A.: How to schedule a cascade in an arbitrary graph. In: EC 2012, pp. 355–368 (2012)
12. Cormode, G., Hadjieleftheriou, M.: Finding frequent items in data streams. Proc. VLDB Endow. 1(2), 1530–1541 (2008)
13. Cormode, G., Muthukrishnan, S.: What's Hot and What's Not: Tracking Most Frequent Items Dynamically. TODS 2005 30(1), 249–278 (2005)
14. The curse of big data,
    http://www.analyticbridge.com/profiles/blogs/the-curse-of-big-data
15. Friedkin, N.E.: The attitude-behavior linkage in behavioral cascades. Social Psychology Quarterly, 73–196 (2010)
16. Glorot, X., Bordes, A., Bengio, Y.: Domain adaptation for large-scale sentiment classification: A deep learning approach. In: ICML 2011 (2011)
17. Hartline, J., Mirrokni, V., Sundararajan, M.: Optimal marketing strategies over social networks. In: WWW 2008, pp. 189–198 (2008)
18. Hong, L., Ahmed, A., Gurumurthy, S., Smola, A.J., Tsioutsiouliklis, K.: Discovering geographical topics in the twitter stream. In: WWW 2012, pp. 769–778 (2012)
19. Horrigan, J., Rainie, L.: When facing a tough decision, 60 million americans now seek the internet's help: The internet's growing role in life's major moments (2006), http://pewresearch.org/obdeck/?ObDeckID=19 (retrieved October 13, 2006)
20. Howe, J.: The rise of crowdsourcing. North 14(14), 1–5 (2006)
21. Hughes, A.L., Palen, L.: Twitter adoption and use in mass convergence and emergency events. International Journal of Emergency Management 6(3/4), 248 (2009)
22. Indonesia, brazil and venezuela lead global surge in twitter usage,
    http://www.comscore.com/Press_Events/Press_Releases/2010/8/
    Indonesia_Brazil_and_Venezuela_Lead_Global_Surge_in_Twitter_Usage

23. Jiang, L., Yu, M., Zhou, M., Liu, X., Zhao, T.: Target-dependent twitter sentiment classification. In: HLT 2011, pp. 151–160 (2011)
24. Jin, C., Qian, W., Sha, C., Yu, J.X., Zhou, A.: Dynamically maintaining frequent items over a data stream. In: CIKM 2003, pp. 287–294. ACM (2003)
25. Katz, I., Tunstrom, K., Ioannou, C., Huepe, C., Couzin, I.: Inferring the structure and dynamics of interactions in schooling fish. In: PNAS 2011, pp. 18720–18725 (2011)
26. Kempe, D., Kleinber, J., Tardos, E.: Maximizing the spread of influence through a social network. In: KDD 2003, pp. 137–146 (2003)
27. Kempe, D., Kleinberg, J.M., Tardos, É.: Maximizing the spread of influence through a social network. In: Proceedings of the Ninth ACM International Conference on Knowledge Discovery and Data Mining, pp. 137–146 (2003)
28. Kifer, D., Ben-David, S., Gehrke, J.: Detecting change in data streams. In: Proc. 30th Int. Conf. on Very Large Data Bases, pp. 180–191. VLDB Endowment (2004)
29. Kimura, M., Saito, K., Nakano, R., Motoda, H.: Extracting influential nodes on a social network for information diffusion. Data Mining and Knowledge Discovery 20, 70–97 (2010)
30. Kimura, M., Saito, K., Ohara, K., Motoda, H.: Learning to predict opinion share in social networks. In: AAAI 2010, pp. 1364–1370 (2010)
31. Kittur, A., Kraut, R.E.: Harnessing the wisdom of crowds in wikipedia: quality through coordination. In: Proceedings of the 2008 ACM Conference on Computer Supported Cooperative Work, CSCW 2008, pp. 37–46. ACM, New York (2008)
32. Krishnamurthy, B., Gill, P., Arlitt, M.: A few chirps about twitter. In: Proceedings of the First Workshop on Online Social Networks, WOSN 2008, pp. 19–24. ACM (2008)
33. Kwak, H., Lee, C., Park, H., Moon, S.: What is Twitter, a social network or a news media. In: WWW 2010, pp. 591–600 (2010)
34. Leskovec, J., Backstrom, L., Kleinberg, J.: Meme-tracking and the dynamics of the news cycle. In: Proc. 15th ACM SIGKDD Int. Conf. on Knowledge Discovery and Data Mining, pp. 497–506 (2009)
35. Leskovec, J., Backstrom, L., Kleinberg, J.: Meme-tracking and the dynamics of the news cycle. In: KDD 2009, pp. 497–506 (2009)
36. Libert, B., Spector, J.: We are smarter than me: how to unleash the power of crowds in your business, 1st edn. Wharton School Publishing (2007)
37. Lin, Y.R., Margolin, D., Keegan, B., Lazer, D.: Voices of Victory: A Computational Focus Group Framework for Tracking Opinion Shift in Real Time. In: WWW 2013, pp. 737–747 (2013)
38. MacEachren, A.M., Robinson, A.C., Jaiswal, A., Pezanov, S., Savelyev, A., Blanford, J., Mitra, P.: Geo-Twitter analytics: Application in crisis management. In: 25th International Cartographic Conference (July 2011)
39. Macropol, K., Singh, A.K.: Content-based modeling and prediction of information dissemination. In: ASONAM 2011, pp. 21–28 (2011)
40. Manku, G.S., Motwani, R.: Approximate frequency counts over data streams. In: VLDB 2002, pp. 346–357 (2002)
41. Mehta, R., Mehta, D., Chheda, D., Shah, C., Chawan, P.: Sentiment analysis and influence tracking using twitter. International Journal of Advanced Research in Computer Science and Electronics Engineering 1, 72–79 (2012)
42. Melville, W.G.P., Lawrence, R.D.: Sentiment analysis of blogs by combining lexical knowledge with text classification. In: KDD 2009, pp. 1275–1284 (2009)

43. Metwally, A., Agrawal, D., El Abbadi, A.: An integrated efficient solution for computing frequent and top-k elements in data streams. ACM Trans. Database Syst. 31(3), 1095–1133 (2006)

44. Metwally, A., Emekçi, F., Agrawal, D., El Abbadi, A.: Sleuth: Single-publisher attack detection using correlation hunting. Proc. VLDB Endow. 1(2), 1217–1228 (2008)

45. Mudhakar, S., Srivatsa, L., Abdelzaher, T.: Mining diverse opinions. In: MILCOM 2012, pp. 1–7 (2012)

46. Palen, L.: Online social media in crisis events. Educause Quarterly (3), 76–78 (2008)

47. Patterson, S., Bamieh, B.: Interaction-driven opinion dynamics in online social networks. In: SOMA 2010, pp. 98–105 (2010)

48. Petrovic, S., Osborne, M., McCreadie, R., Macdonald, C., Ounis, I., Shrimpton, L.: Can Twitter replace Newswire for breaking news? In: ICWSM 2013, pp. 713–716 (2013)

49. Rosenfeld, A., Hummel, R.A., Zucker, S.W.: Scene labeling by relaxation operations. IEEE Transactions on Systems Man and Cybernetics 6, 420–433 (1976)

50. Sachan, M., Contractor, D., Faruquie, T.A., Subramaniam, L.V.: Using content and interactions for discovering communities in social networks. In: WWW 2012, pp. 331–340 (2012)

51. Sakaki, T., Okazaki, M., Matsuo, Y.: Earthquake shakes twitter users: real-time event detection by social sensors. In: Proceedings of the 19th International Conference on World Wide Web, WWW 2010, pp. 851–860. ACM, New York (2010)

52. Sankaranarayanan, J., Samet, H., Teitler, B.E., Lieberman, M.D., Sperling, J.: Twitterstand: news in tweets. In: GIS 2009: Proceedings of the 17th ACM SIGSPATIAL International Conference on Advances in Geographic Information Systems, pp. 42–51. ACM, New York (2009)

53. Teitler, B.E., Lieberman, M.D., Panozzo, D., Sankaranarayanan, J., Samet, H., Sperling, J.: Newsstand: a new view on news. In: GIS 2008, pp. 1–10 (2008)

54. Teitler, B.E., Lieberman, M.D., Panozzo, D., Sankaranarayanan, J., Samet, H., Sperling, J.: Newsstand: a new view on news. In: GIS 2008: Proceedings of the 16th ACM SIGSPATIAL International Conference on Advances in Geographic Information Systems, pp. 1–10. ACM, New York (2008)

55. Thelwall, M., Buckley, K., Paltoglou, G., Cai, D., Kappas, A.: Sentiment strength detection in short informal text. Journal of the American Society for Information Science and Technology 61, 2544–2558 (2010)

56. Twitter, http://www.twitter.com

57. Wang, X., Wei, F., Liu, X., Zhou, M., Zhang, M.: Topic sentiment analysis in twitter: a graph-based hashtag sentiment classification approach. In: CIKM 2011, pp. 1031–1040 (2011)

58. Wu, M.: The big data fallacy and why we need to collect even bigger data. TechCrunch (2012)

An Inductive and Semantic Model of Constraints for Master Data Management under Cloud Computing

Shinji Kikuchi

School of Computer Science and Engineering, University of Aizu,
Ikki-machi, Aizu-Wakamatsu City, Fukushima 965-8580, Japan
shinji.kikuchi@ieee.org

Abstract. Master Data Management (MDM) has been evaluated under the contexts of Enterprise Architecture (EA), Semantic Web, Service Oriented Architecture (SOA) and Business Process Integration (BPI). However, there have been very few studies on operations of MDM under a Cloud Computing environment. Accordingly, we have analyzed the operational issues caused by migrating the MDM operations into the Cloud Computing environment, in particular into Software as a Service. Furthermore, we have pointed out the insufficient policy of the security framework in regards to the disclosure, and also pointed out the necessity to define a constraints model for security policy that is available in designing the operations of MDM among multiple SaaS/ASP and BPaaS providers. In this paper, we define a more precise constraints model by applying the first order predicate logic and inductive approach. Furthermore, we give explanatory description about the contexts of this formularization.

Keywords: Master Data Management, Constraints Model, Cloud Computing, Formularization.

1 Introduction

Enterprise Application Integration (EAI) for an enterprise and Business Process Integration (BPI) over multiple independent enterprises are carried out. The normalization and standardization on business data have been evolved [1]. In particular, the importance of Master Data Management (MDM) has also been highlighted [1], [2]. Accordingly, we have analyzed the operational issues caused by migration of the MDM operations to a Cloud Computing environment, in particular to Software as a Service (SaaS) [3]. This trend is important since Business Process as a Service (BPaaS) has emerged [4]. For instance, Fig.1 shows our proposed model as the typical environment and also explains about a simplified issue. Here, we assume multiple users named as Enterprise.x (x: 1, 2,....,n) exist, and multiple providers named as SaaS/ASP.y (y: 1, 2,....,m) exist as well. Then, inside Enterprise.1 a MDM is implemented for local uses and named as 'Local MDM'. On the other hand, inside SaaS/ASP.1, another MDM managed for a commercial service and providing it to others is also implemented. We can regard this as 'Global MDM', because this can handle the demands from multiple users as multi-tenancies. Global MDM can provide services to multiple Enterprise.x (x: 1, 2,....,n). Furthermore, we also assume UDDI and other repositories (UDDI/Repository) will be deployed inside SaaS/ASP.m.

A. Madaan, S. Kikuchi, and S. Bhalla (Eds.): DNIS 2014, LNCS 8381, pp. 17–39, 2014.

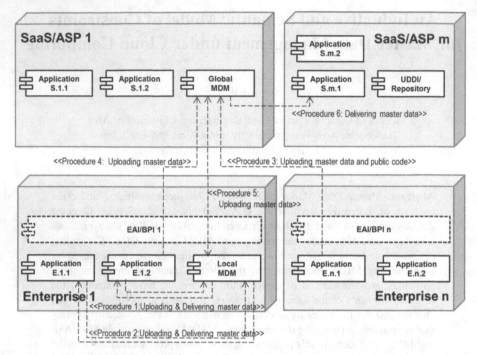

Fig. 1. Workflow inside an enterprise under Cloud environment using MDM procedures

The procedures for exchanging master data will be executed as follows; Inside Enterprise.1, all of the application programs will update the master data which is managed by themselves locally by sending them to Local MDM. Procedure.1 and Procedure.2 will be activated in order to normalize these master data. Then data cleansing and normalizing these master data will be carried out. Delivering normalized master data coupled with the original master data to the original sender applications are continuously performed. Through these processes, semantics integration with uniformed expressions is realized among the master data. In the case of utilizing the Global MDM service at SaaS/ASP.1 by multiple Enterprise.x (x:1,2,....,n) , these Enterprise.x (x:1,2,....,n) must send their master data to Global MDM first. Enterprise.n for example, uploads all of its master data to Global MDM at Proccdure.3, then data cleansing will be started. In this case, the disclosed data to the outside by Enterprise.n such as identification information are also uploaded at that time. Enterprise.1 uploads all of its master data to Global MDM as well at Procedure.4 and 5. After that data cleansing will be done, although an operational rule must be established, which Global MDM at SaaS/ASP.1 or Local MDM managed by Enterprise.1 should be prioritized. If Enterprise.1 prioritizes Global MDM at SaaS/ASP.1, Enterprise.1 must deliver the normalized master data from Global MDM to another SaaS/ASP.m at Procedure.6. Although this procedure can be regarded as a proxy activity, it might cause an issue on the authentication mechanism. In this case the service provider SaaS/ASP.1 might know the fact about the contract of its client Enterprise.l to another service provider. In order to avoid this solution, it should be required to develop a method to make multiple service providers invisible to each other. This will potentially become an operational constraint for MDM.

Except the above issue, there are naturally several potential issues. We have also pointed out these operational issues that would potentially occur [3]. Furthermore, we mentioned that there are some dependencies among these issues and independencies from each other. The relationships among them could be depicted as Fig.2. [3] Based on observed independencies, we categorized them into two of the following major groups; the first group is related to the matters of the operation of UDDI as the registry system. Conversely, the second group is related to the disclosure of the master data within the contracted group. This group contains the most of the connected issues regarded as dependencies. Furthermore, the second group also contains an issue of how to map the sub-solutions into the given security framework. Linked with this second group, we also introduced the formularized framework of the constraints, which is aimed to fill the lack of descriptions as the security requirement, despite remaining at an immature level [3].

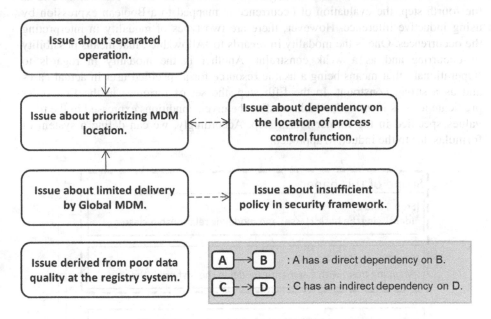

Fig. 2. Dependencies among identified issues

Based on the previous work, in this paper, we define a more precise constraints model with applying the first order predicate logic and inductive approach. In particular, we describe the detail explanation of this formularization together.

As for the formalization process, we should ideally take the procedure depicted as Fig.3. In this approach, at first we should grasp a set of essential classes and relationship-classes, then we should continuously make a semantic model. After that, we need to prove the consistency and completeness by first order predicate calculus after defining the syntax with the set of the logical axioms. This process should recursively be carried out until proving the completeness of first order theorem. In this paper, we will particularly focus on defining the semantic model.

During the procedures of defining the semantic model, the main stream consists of five steps that Table.1 shows. The first step is to identify the set of the most basic classes and to define them. Then, the individual class is assigned to a set which has finite elements. The second step is to extract the set of the relationship-classes linked with the above basic classes. Occurrence of each instance of each relationship-class corresponds to arising of an actual phenomenon; therefore it is mapped to a Boolean expression. There is obviously a constraint in regards to co-occurrences among instances of the relationship-classes. Thus, as the third step, we extract the relationships, which are decided by the combinations consisting of the individual instances of the relationship-classes. However, this extraction process is done in the inductive approach in which the scope of the relationships should be decided with the actual or probable occurrences rather than all of the conceptually possible cases. After that, the extracted relationships are regarded as candidates' set of constraints, and as the fourth step, the evaluation of occurrence is mapped to a Boolean expression by using inductive inference. However, there are two types of modality in interpreting the occurrences. One is the modality in regards to 'Allowable', that means possibility of occurring and as a weak constraint. Another is the modality in regards to 'Operational', that means being a usable resource for a specified user in actual cases and as a strong constraint. In the fifth step, the set of formulas in the first order predicate logic is identified, which satisfies the given conditions expressed in the truth values specified in the former fourth step. Accordingly, we can define a system of formulas due to the inductive approach.

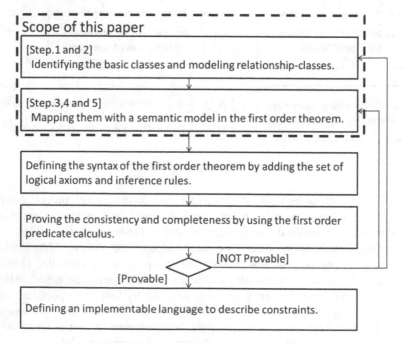

Fig. 3. Taken formularization process

The remainder of this is organized as follows; in section 2, we define the set of the most basic classes clarified in step.1 and that of the relationship-classes extracted in step.2. After that we also define the basic essential set of predicates. These predicates are applied and used in evaluation as the truth table. In section 3, we mention step.3 with extracting the relationships inductively, which are decided by the combinations consisting of the individual predicates. Additionally, we also talk about the set of the bases of inductive mapping between the evaluation of an occurrence and the Boolean expression in step.4. This is the most crucial factor to define the constraints. In section 4, as the results of step.5, we list the set of the identified formulas in the first order predicate logic, which satisfies the given conditions expressed in the truth tables extracted in section 3. However, the set of proof about completeness without any contradictions inside this model will not be included, here. In section 5, as our considerations, we briefly mention the appropriateness and practical usages of this model and then also summarize the related works. Finally in section 6, we conclude and discuss future's direction.

Table 1. Five steps to define a semantic model

Step #	Definitions
Step.1	Identifying the set of the most basic classes.
Step.2	Extracting and defining the set of the relationship-classes among the basic classes.
Step.3	Identifying the potential relationships by combining individual predicates and considering set of the co-occurrences of them.
Step.4	Evaluating occurrences of the potential relationships by using inductive inference based on the actual or the potential constraints.
Step.5	Identifying a semantic model by giving the set of formulas in the first order predicate logic, that can satisfy the evaluation results given at the Step.4.

2   Definition of the Logical Framework

Fig.4 shows the class diagram of the most essential part of areas we are focusing on. The most primitive classes are 'User' regardless of an ownership, 'Master Data' including both definition of the type and its instances, and 'Site' meaning providers. The class 'User' will be specialized into the classes 'Owner', or 'Non-owner'. Instances of the class 'Master Data' will be commonly applicable even if a cleansing process is finished over them. However, we can regard that they roughly have one of the following two statuses, as 'Original' and 'Cleansed'. Accordingly, as our initial preparation, we can define several sets as follows;

i) $\{M_o\}$ expresses a set of original instances of a master data.
ii) $\{M_c\}$ expresses a set of cleansed instances of the master data corresponding to the above $\{M_o\}$.
iii) $\{U\}$ expresses a set of potential users regardless of types.
iv) $\{S\}$ expresses a set of instances of site.

On the other hand, there are five relationships as following; the first is 'Contract' which means the contract between a user instance and a site instance. The second is 'Own' meaning the ownership between a user instance and an instance of the master data. The third is 'Disclose' which means permission whether a user instance can refer a specified instance of the master data. The fourth is 'Deploy' meaning the status whether a specified instance of the master data is deployed to a particular site. And the final relationship is 'Delivery' which deals with the information whether the master data could be shared or not. However this 'Delivery' will be available only under the multiple Cloud environments. The original site of the transferred master data will be expressed with the suffix 's' which means 'Source'. Whereas, the receiving site of that will be expressed with the suffix 'd' which means 'Destination'. In the first order predicate logic, we can map these relationships with set of corresponding predicates in the following expressions (1), (2), (3), (4) and (5):

$$\exists u (u \in \{U\}), \exists s (s \in \{S\}) \, Con(u,s) \tag{1}$$
$$\exists u (u \in \{U\}), \exists m (m \in \{M_o\} \cup \{M_c\}) \, Own(u,m) \tag{2}$$
$$\exists m (m \in \{M_o\} \cup \{M_c\}), \exists u \, (u \in \{U\}) \, Dis(m,u) \tag{3}$$
$$\exists m (m \in \{M_o\} \cup \{M_c\}), \exists s (s \in \{S\}) \, Dep(m,s) \tag{4}$$
$$\exists m (m \in \{M_o\} \cup \{M_c\}), \exists s_s, \exists s_d (s_s, s_d \in \{S\}, s_s \neq s_d) \, Dlv(m, s_s, s_d) \tag{5}$$

Furthermore, we can define other sets as follows;

v)   {C} expresses a set of any general common formulas, which contains the constrains that are defined with previous sets of $\{M_o\}$, $\{M_c\}$, {U}, {S} and set of the predicates of 'Contract', 'Own', 'Disclose' and 'Deploy' as expressions (1),(2),(3) and (4).

vi) {C_{cloud}} expresses a set of formulas, which contains constrains that are defined for the Cloud environments only. In this case, the predicate of 'Delivery' as expression (5) will be contained.

Therefore, when we treat an issue around the Cloud environments, the set of the constrains should be defined as a following union; {C} $\cup$ {C_{cloud}}. Here, we define the first order language of the system we are concerned about in Fig.4. The symbol {⟨X⟩} could be regarded as the set of the symbols which correspond to the individual elements of a set X. Conversely, the symbol {Meta} could be regarded as the set of the symbolic items such as variables, constants, terms of the meta language making the first order predicate logic. In particular, a pair of symbols {⊤, ⊥} corresponds to that of 'True' and 'False'. Our expression in the first order language is defined as (6), and an interpretation I could be described as (7).

$$(\{\langle M_o \rangle\} \cup \{\langle M_c \rangle\} \cup \{\langle U \rangle\} \cup \{\langle S \rangle\} \cup \{Meta\}, \{C\} \cup \{C_{cloud}\}) \tag{6}$$
$$(\{M_o\} \cup \{M_c\} \cup \{U\} \cup \{S\}, \mu, \rho, \varphi) \tag{7}$$

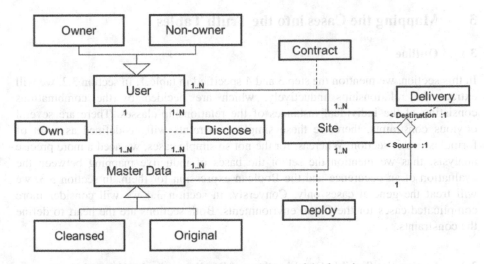

Fig. 4. Class diagram of the essential part of the focusing area

In the above (7), μ means a mapping in which the individual constants in the set $\{\langle M_o\rangle\} \cup \{\langle M_c\rangle\} \cup \{\langle U\rangle\} \cup \{\langle S\rangle\}$ correspond to the entities of our target domains. ρ means the other mapping in which the individual functions having n-ary in the same set correspond to the specified functions having the same n-ary of our target domains. On the other hand, φ means the other mapping in which the individual predicates having n-ary in the same set correspond to the specified predicates having the same n-ary of our target domains. If an expression consisting of formula as f is true under an interpretation $\mathbf{I}$, we define this satisfaction by using the expression (8).

$$\llbracket f \rrbracket^I = \top \;\leftrightarrow\; \vDash_I f \tag{8}$$

In our semantic, there are multiple modalities when we evaluate the formulas with the truth tables. In our model we treat these modalities as the corresponding interpretations. The first interpretation is 'Allowable', which means that it is not always impossible. In detail, this involves that our hypothetical situations will potentially occur despite the large potential result of an operational failure. On the other hand, the second interpretation is 'Operational', in which any conditions where our system does not accept for the operations should be excluded. Furthermore, there is an intermediate interpretation named as 'Available' which is not equivalent with 'Operational', but results in becoming operative. If we express the strength of the constraints of a modality $\mathbf{X}$ as $\|\mathbf{X}\|$ we could get the following order in (9). The left side in the order is called as 'weak constraint', whereas the right side is called as 'strong constraint'. Furthermore (10) could be satisfied as a logical consequence.

$$\|Allowable\| \;\ll\; \|Available\| \;\ll\; \|Operational\| \tag{9}$$

$$\forall f\left(\vDash_{Operational} f \;\rightarrow\; \vDash_{Allowable} f\right) \tag{10}$$

3   Mapping the Cases into the Truth Tables

3.1   Outline

In this section, we mention the step.3 and 4 specified in table.1. In section 3.2, we will extract the relationships inductively, which are decided by the combinations consisting of the individual instances of the relationship-classes. There are several obvious constraints, therefore these simple constraints will be defined as a set of formulas in this section. Whereas, for the not so simple cases, we need a more precise analysis, thus we mention the set of the bases of inductive mapping between the evaluation of an occurence and the Boolean expression for them. In section 3.3, we will treat the general cases only. Conversely, in section 3.4 we will consider more complicated cases for the Cloud envrionments. Both sections are the heart to define the constraints.

3.2   Listing the Possible Combinations of Predicates in the General Cases

Table.2 defines the cases of possible and meaningful combinations among predicates used in the general common formulas in order to identify the constraints. There are seven potential combinations and they are individually named from 'Possibility-1' to 'Possibility-7'. By evaluating all of these possibilities one by one, we inductively try to identify the relationships as constraints.

Table 2. Possible combinations among predicates

	Own(u,m)	Dis(m,u)	Dep(m,s)	Con(u,s)
Own(u,m)	(1) Possibility-1	-	-	-
Dis(m,u)	(2) Possibility-2	(3) Possibility-4	-	-
Dep(m,s)	(4)	(5)	(6) Possibility-6	-
Con(u,s)	(7) Possibility-3	(8) Possibility-5	(9)	(10) Possibility-7

Possibility-1 is related to the constraints about the double ownerships. This means that an ownership on a master data must be identified without any duplication. Accordingly, the formula expressed as (13) is defined, if the predicate in regards to identification of objects is supplementally given as the paired expressions of (11) and (12). This formula (13) means identifiable without any duplication. Furthermore, another formula expressed as (14) is also definable. This formula (14) means a constraint about no master data without any ownership.

$$\vDash_{\{Allowable,Available,Operational\}} \forall x \big(x \in (\{M_o\} \cup \{M_c\} \cup \{U\} \cup \{S\})\big) \, Idy(x,x) \qquad (11)$$

$$\vDash_{\{Allowable,Available,Operational\}} \forall x, \nexists x'(x,x' \in (\{M_o\} \cup \{M_c\} \cup \{U\} \cup \{S\}), x \neq x')$$
$$Idy(x,x') \qquad (12)$$

$$\vDash_{\{Allowable,Available,Operational\}} \forall u(u \in \{U\}), \exists m(m \in \{M_o\} \cup \{M_C\})$$
$$(Own(u,m) \rightarrow \nexists u' \, (u' \in \{U\}, u' \neq u)(Own(u',m) \wedge Idy(u',u))) \qquad (13)$$

$$\vDash_{\{Allowable,Available,Operational\}} \nexists u(u \in \{U\}), \forall m(m \in \{M_o\} \cup \{M_C\}) \neg Own(u,m)(14)$$

When classifying the states between pre cleansing of the master data and its post cleansing, there is another constraint and will be defined later. Possibility-2 is a relationship between ownership and 'Disclose'. There is an obvious constraint in this relationship. The formula expressed as (15) means that all of the master data must naturally be disclosed to the owner of them. Furthermore this formula (15) involves the case where a master data might also be allowable to be disclosed to users who are not owners of that master data.

$$\vDash_{\{Allowable,Available,Operational\}} \forall u(u \in \{U\}), \exists m(m \in \{M_o\} \cup \{M_C\})$$
$$(Own(u,m) \rightarrow Dis(m,u)) \qquad (15)$$

The relationship between 'Deploy' and 'Contract' is treated in Possibility-3 and 5. These Possibility-3 and 5 individually correspond to the constraints caused by two independent relationships among three relationship-classes. They will be explained in section 3.3. When neglecting a state difference between pre cleansing of the master data and its post cleansing, there are no constraints related to the ownership in the case of Possibility-4, and this is different from that of the previous Possibility-1. When specifying a master data under the condition of multiple existing users, the result of the predicate 'Disclose' will merely derive from the access permissions as authorization. Conversely, when specifying a user under the condition where multiple related master data exist, the accessible master data will be decided dependently on the specified access permissions as authorization. Therefore, there are no constraints except that derived from the above authorization.

When neglecting a state difference between pre cleansing of the master data and its post cleansing, there are no relationships and constraints between a master data and a site in the case of Possibility-6. This is obvious because it is possible to deploy the replicas of the specified master data on multiple sites than a single one. Furthermore, it is also possible to manage any number of multiple master data including zero or one. Therefore, there are no rooms where any constraints occur.

Possibility-7 corresponds to the relationships among multiple contracts made between a site and a user. As far as this simplified model is concerned, there are no constraints. In general, a single user can individually make contracts with multiple sites including zero. And a site will have a set of contracts with multiple users. If there would be a constraint, it would take place when we would extend our current model with other added classes.

When classifying the states between pre cleansing of the master data and its post cleansing, there are additional constraints for Possibility-1, 4 and 6 individually. As for the Possibility-1, a new constraint expressed as the formula (16) should be added. This means that there are no changes around the ownership of the specified master data between the both states.

$$\vDash_{\{Allowable,Available,Operational\}} \forall u(u \in \{\mathbf{U}\}), \forall m(m \in \{\mathbf{M_O}\})$$
$$(Own(u,m) \longleftrightarrow \exists|m|(|m| \in \{\mathbf{M_C}\})Own(u,|m|)) \quad (16)$$

As for the Possibility-4, a new constraint expressed as the formula (17) should also be added. This means that there are no changes about 'Disclose' of the specified master data between the both states.

$$\vDash_{\{Allowable,Available,Operational\}} \forall m(m \in \{\mathbf{M_O}\}), \exists u(u \in \{\mathbf{U}\})$$
$$(Dis(m,u) \longleftrightarrow \exists|m|(|m| \in \{\mathbf{M_C}\})Dis(|m|,u)) \quad (17)$$

In regards to the Possibility-6, some constraints with respect to the cleansing are considerable. These constraints related to 'Deployment' might be variable and depend on the states between pre cleansing of the master data and its post cleansing, despite several causing factors. However, we will evaluate them according to the types of modality in interpretting the occurences 'Allowable'or 'Operational' instead of analyzing the factors in detail. Table.3 shows the desired results of evaluated cases about the dependency of deployment on the states between pre cleansing of the master data and its post cleansing. In the case of interpretation of 'Allowable', the case of (2) is probable because there might be some delays in an actual deployment. On the other hand, the case of (3) is also possible in practice for example, because of a backup of data carried out after the cleansing. Therefore we should regard this as a probable case. As a conclusion, under the interpretation of 'Allowable', this should be treated as always true. Conversely, in the case of interpretation of 'Operational', the cases of (2) and (3) will be assumed not to be permitted, for example due to the security reasons.

Table 3. Dependency of deployment on cleansing

Cases	$Dep(m,s)$	$Dep(|m|,s)$	Desired Result (Reason) under Allowable cases.		Desired Result (Reason) under Operational cases.	
(1)	T	T	T	Obvious.	T	Obvious.
(2)	T	⊥	T	Because there might be some delay in deployment.	⊥	Not allowed.
(3)	⊥	T	T	Because the specified site might be a backup.	⊥	Not preferable due to security reasons.
(4)	⊥	⊥	T	Obvious.	T	Obvious.

Based on the table.3, the formulas (18), (19) expressing the above constraints should be added.

$$\vDash_{Allowable} \forall m(m \in \{\mathbf{M_O}\}), \exists s(s \in \{\mathbf{S}\})$$
$$(Dep(m,s) \rightarrow \exists|m|(|m| \in \{\mathbf{M_C}\})(\neg Dep(|m|,s) \vee Dep(|m|,s))) \quad (18)$$
$$\vDash_{Operational} \forall m(m \in \{\mathbf{M_O}\}), \exists s(s \in \{\mathbf{S}\})$$
$$(Dep(m,s) \longleftrightarrow \exists|m|(|m| \in \{\mathbf{M_C}\})Dep(|m|,s)) \quad (19)$$

3.3   Evaluation of an Occurrence in the General Cases

In this section, we mention the set of the bases of inductive mapping between the evaluation of an occurence and the Boolean expression for the previous two possiblity-3 and 5. In each possibility, we evaluate cases and specify the desired results. Therefore, the set of derived formulas must satisfy these desired results as mentioned in the following section. Table.4 shows the desired results of evaluated cases about the co-occurrence among the previous three predicates 'Own', 'Deploy' and 'Contract'. As mentioned previously, the desired results will be variable with dependency on the modality. Allowable cases correspond to the interpretation of 'Allowable'. In these cases, according to the values which the three predicates 'Own', 'Deploy' and 'Contract' individually have, the obvious impossible cases are evaluated as false. Whereas, the possible cases regardless of operational conditions are evaluated as true. Case (2) is false because this means the master data is already deployed in spite of no contracts. Case (5) is true because there are several possibilities that some access permissions as authorization might be positively specified in spite of another user's ownership of the specified master data. Case (6) is also true because there are several possibilities that the specified user does not have any relationships to the specified master data. The other cases except the cases (2), (5) and (6) are obvious. There is another modality that correspunds to the interpretation of 'Operational'. Under this interpretation, case (1) is the only allowable case. As for the case (2), it should be false because of the same reason of the interpretation of 'Allowable'. On the other hand, the case (5) would be 'Cannot be decided' because of the dependency on access permissions as authorization. The other cases except the cases (2) and (5) are obvious because of no contracts or no deployed master data.

Table 4. Evaluation among predicates 'Own', 'Deploy' and 'Contract'

Cases	Own(u,m)	Dep(m,s)	Con(u,s)	Desired Result (Reason) under Allowable cases.		Desired Result (Reason) under Operational cases.	
(1)	T	T	T	T	Obvious.	T	Obvious.
(2)	T	T	$\perp$	$\perp$	Because the specified master is deployed without a contract.	$\perp$	Because the specified master is deployed without a contract.
(3)	T	$\perp$	T	T	Obvious.	$\perp$	Obvious.
(4)	T	$\perp$	$\perp$	T	Obvious.	$\perp$	Obvious.
(5)	$\perp$	T	T	T	Owned by another user.	Cannot be decided	Because the result depends on the given authorization to the specified user.
(6)	$\perp$	T	$\perp$	T	Because the deployed master might be owned by another user.	$\perp$	Obvious.
(7)	$\perp$	$\perp$	T	T	Obvious.	$\perp$	Obvious.
(8)	$\perp$	$\perp$	$\perp$	T	Obvious.	$\perp$	Obvious.

Table.5 shows the desired results of evaluated cases about the co-occurrence among the previous three predicates 'Disclose','Deploy' and 'Contract'. As mentioned previously, the desired results will be variable with dependency on the modality. Allowable cases correspond to the interpretation of 'Allowable', and regardless of the combinations of the values which the three predicates 'Disclose','Deploy' and 'Contract' individually have, all cases are evaluated as true. Case (2) is also true. In this case, there are several possibilities that the specified master data might be deployed to another site, although the specified user does not have any contract with the specified site. However, it becomes false under the interpretation of 'Operational' because there are obviously certain difficult cases for the actual operation. The reason why the case (5) is true under the interpretation of 'Allowable' is obvious when considering the particular case of the ownership by another user. There are several possibilities that access permissions as authorization are not specified positively in spite of deployment of the specified master data. However, it also becomes false under the interpretation of 'Operational' because there are obviously certain difficult cases for the actual operation as well as the previous case (2). The reason why the case (6) is true under the interpretation of 'Allowable' is that there are several possibilities that the specified and deployed master data might not be needed for the specified user. However, under the interpretation of 'Operational' it should obviously be false because of no contract. The other cases except the cases (2), (5) and (6) are obviously true, in particular under the interpretation of 'Allowable'. Whereas, they are obviously false because of no contracts or no deployed master data.

Table 5. Evaluation among predicates 'Disclose','Deploy' and 'Contract'

Cases	Dis(m,u)	Dep(m,s)	Con(u,s)	Desired Result (Reason) under Allowable cases.		Desired Result (Reason) under Operational cases.	
(1)	T	T	T	T	Obvious.	T	Obvious.
(2)	T	T	⊥	T	Because the master might be accessible by the specified user at another site.	⊥	No Contracts.
(3)	T	⊥	T	T	Obvious.	⊥	Obvious.
(4)	T	⊥	⊥	T	Obvious.	⊥	Obvious.
(5)	⊥	T	T	T	Obvious.	⊥	Because the master is not disclosed to the specified user.
(6)	⊥	T	⊥	T	Because the deployed master might not be needed.	⊥	No Contracts.
(7)	⊥	⊥	T	T	Obvious.	⊥	Obvious.
(8)	⊥	⊥	⊥	T	Obvious.	⊥	Obvious.

Table 6. List of the reasons

Symbol	Definitions
[R#1]	Conflict due to being deliverable without a contract.
[R#2]	No conflict due to being not deliverable without a contract.
[R#3]	Conflict due to an existing master data without a contract.
[R#4]	Conflict due to a pre-deployed master data.
[R#5]	Possible to obtain.
[R#6]	Depends on the given authorization to the specified user.
[R#7]	Possible to access.
[R#8]	No source master data.
[R#9]	Impossible operation due to being not deliverable.

3.4   Evaluation of an Occurrence in the Particular Cases of Cloud Computing

In this section, we mention the set of the bases of inductive mapping between the evaluation of an occurence and the Boolean expression for more complicated cases in the cloud envrionments. In this case, there is the third interpretation named as 'Available' corresponding to the modality except two mentioned interpretations of the general cases in sections 3.2 and 3.3. This means the available master data which is already deployed at the destination site for whatever reasons. The constraints under this interpretation would be weakened than that of interpretation 'Operational'. Furthermore, there is an obvious dependency of desired results on a state difference between pre delivery and post-delivery. For instance, if there would be a deployed master data in spite of being banned to deliver it to the destination side previously, the desired result should be specified as false because of an operational conflict. Tables.7, 8, 9 and 10 show the desired results of evaluated cases about the co-occurrence among multiple predicates including 'Delivery'. In this case, the following six predicates are related, 'Delivery', 'Own', 'Deploy' and 'Contract' of both source and destination sides. Table.6 shows the definitions of the symbols and their meanings in regards to the reasons for the desired results arisen in the tables 7, 8, 9 and 10. The desired results in the tables 7, 8, 9 and 10 derive from combinations of multiple explanatory reasons in the table.6.

Table 7. Evaluation of cases with predicate 'Delivery' (1/4)

Desired Result (Reasons) under Allowable cases without the initial master data at the destination site.	Desired Result (Reasons) under Allowable cases with the initial master data at the destination site.	Desired Result (Reasons) under Operational cases without the initial master data at the destination site.	Desired Result (Reasons) under Operational cases with the initial master data at the destination site.	Own (u,m)	Dep (m,s_o)	Con (u,s_o)	Dep (m,s_d)	Con (u,s_d)	Div (m,s_o,s_d)
⊥ [R#4]	T	⊥ [R#4]	T	T	T	T	T	T	T
⊥ [R#4]	T	⊥ [R#4][R#9]	⊥ [R#9]	T	T	T	T	T	⊥
⊥ [R#1][R#3][R#4]	⊥ [R#1][R#3]	⊥ [R#1][R#3][R#4]	⊥ [R#1][R#3]	T	T	T	T	⊥	T
⊥ [R#3][R#4]	⊥ [R#3]	⊥ [R#3][R#4][R#9]	⊥ [R#3][R#9]	T	T	T	T	⊥	⊥
T	T	T	T	T	T	T	⊥	T	T
T	T	⊥ [R#9]	⊥ [R#9]	T	T	T	⊥	T	⊥
⊥ [R#1]	⊥ [R#1]	⊥ [R#1]	⊥ [R#1]	T	T	T	⊥	⊥	T
T [R#2]	T [R#2]	⊥ [R#9]	⊥ [R#9]	T	T	T	⊥	⊥	⊥
⊥ [R#1][R#3][R#4]	⊥ [R#1][R#3]	⊥ [R#1][R#3][R#4]	⊥ [R#1][R#3]	T	T	⊥	T	T	T
⊥ [R#3][R#4]	⊥ [R#3]	⊥ [R#3][R#4][R#9]	⊥ [R#3][R#9]	T	T	⊥	T	T	⊥
⊥ [R#1][R#3][R#4]	⊥ [R#1][R#3]	⊥ [R#1][R#3][R#4]	⊥ [R#1][R#3]	T	T	⊥	T	⊥	T
⊥ [R#3][R#4]	⊥ [R#3]	⊥ [R#3][R#4][R#9]	⊥ [R#3][R#9]	T	T	⊥	T	⊥	⊥
⊥ [R#1][R#3]	⊥ [R#1][R#3]	⊥ [R#1][R#3]	⊥ [R#1][R#3]	T	T	⊥	⊥	T	T
⊥ [R#1]	⊥ [R#3]	⊥ [R#3][R#9]	⊥ [R#3][R#9]	T	T	⊥	⊥	T	⊥
⊥ [R#1][R#3]	⊥ [R#1][R#3]	⊥ [R#3][R#9]	⊥ [R#3][R#9]	T	T	⊥	⊥	⊥	T
⊥ [R#3]	⊥ [R#3]	⊥ [R#3][R#9]	⊥ [R#3][R#9]	T	T	⊥	⊥	⊥	⊥
⊥ [R#4]	T	⊥ [R#8][R#4]	⊥ [R#8]	T	⊥	T	T	T	T
⊥ [R#4]	T	⊥ [R#8][R#9][R#4]	⊥ [R#8][R#9]	T	⊥	T	T	T	⊥
⊥ [R#1][R#3][R#4]	⊥ [R#1][R#3]	⊥ [R#1][R#3][R#8][R#4]	⊥ [R#1][R#3][R#8]	T	⊥	T	T	⊥	T
⊥ [R#3][R#4]	⊥ [R#3]	⊥ [R#3][R#8][R#9][R#4]	⊥ [R#3][R#8][R#9]	T	⊥	T	T	⊥	⊥
T	T	⊥ [R#8]	⊥ [R#8]	T	⊥	T	⊥	T	T
T	T	⊥ [R#8][R#9]	⊥ [R#8][R#9]	T	⊥	T	⊥	T	⊥
⊥ [R#1]	⊥ [R#1]	⊥ [R#1][R#8]	⊥ [R#1][R#8]	T	⊥	T	⊥	⊥	T
T [R#2]	T [R#2]	⊥ [R#8][R#9]	⊥ [R#8][R#9]	T	⊥	T	⊥	⊥	⊥
⊥ [R#1][R#4]	⊥ [R#1]	⊥ [R#1][R#8][R#4]	⊥ [R#1][R#8]	T	⊥	⊥	T	T	T
⊥ [R#4]	T [R#2]	⊥ [R#8][R#9][R#4]	⊥ [R#8][R#9]	T	⊥	⊥	T	T	⊥
⊥ [R#1][R#3][R#4]	⊥ [R#1][R#3]	⊥ [R#1][R#3][R#8][R#4]	⊥ [R#1][R#3][R#8]	T	⊥	⊥	T	⊥	T
⊥ [R#3][R#4]	⊥ [R#3]	⊥ [R#3][R#8][R#9][R#4]	⊥ [R#3][R#8][R#9]	T	⊥	⊥	T	⊥	⊥
⊥ [R#1]	⊥ [R#1]	⊥ [R#1][R#8]	⊥ [R#1][R#8]	T	⊥	⊥	⊥	T	T
T [R#2]	T [R#2]	⊥ [R#8][R#9]	⊥ [R#8][R#9]	T	⊥	⊥	⊥	T	⊥
⊥ [R#1]	⊥ [R#1]	⊥ [R#1][R#8]	⊥ [R#1][R#8]	T	⊥	⊥	⊥	⊥	T
T [R#2]	T [R#2]	⊥ [R#8][R#9]	⊥ [R#8][R#9]	T	⊥	⊥	⊥	⊥	⊥

Table 8. Evaluation of cases with predicate 'Delivery' (2/4)

Desired Result (Reasons) under Allowable cases without the initial master data at the destination site.	Desired Result (Reasons) under Allowable cases with the initial master data at the destination site.	Desired Result (Reasons) under Operational cases without the initial master data at the destination site.	Desired Result (Reasons) under Operational cases with the initial master data at the destination site.	Own (u,m)	Dep (m,s_o)	Con (u,s_o)	Dep (m,s_d)	Con (u,s_d)	Div (m,s_o,s_d)
⊥ [R#4]	T	⊥ [R#4]	T [R#6]	⊥	T	T	T	T	T
⊥ [R#4]	T	⊥ [R#9][R#4]	⊥ [R#9]	⊥	T	T	T	T	⊥
⊥ [R#1][R#3][R#4]	⊥ [R#1][R#3]	⊥ [R#1][R#3][R#4]	⊥ [R#1][R#3]	⊥	T	T	T	⊥	T
⊥ [R#3][R#4]	⊥ [R#3]	⊥ [R#3][R#4][R#9]	⊥ [R#3][R#9]	⊥	T	T	T	⊥	⊥
T	T	T [R#6]	T [R#6]	⊥	T	T	⊥	T	T
T	T	⊥ [R#9]	⊥ [R#9]	⊥	T	T	⊥	T	⊥
⊥ [R#1]	⊥ [R#1]	⊥ [R#1]	⊥ [R#1]	⊥	T	T	⊥	⊥	T
T [R#2]	T [R#2]	⊥ [R#9]	⊥ [R#9]	⊥	T	T	⊥	⊥	⊥
⊥ [R#1][R#3][R#4]	⊥ [R#1][R#3]	⊥ [R#1][R#3][R#4]	⊥ [R#1][R#3]	⊥	T	⊥	T	T	T
⊥ [R#3][R#4]	⊥ [R#3]	⊥ [R#3][R#4][R#9]	⊥ [R#3][R#9]	⊥	T	⊥	T	T	⊥
⊥ [R#1][R#3][R#4]	⊥ [R#1][R#3]	⊥ [R#1][R#3][R#4]	⊥ [R#1][R#3]	⊥	T	⊥	T	⊥	T
⊥ [R#3][R#4]	⊥ [R#3]	⊥ [R#3][R#4][R#9]	⊥ [R#3][R#9]	⊥	T	⊥	T	⊥	⊥
⊥ [R#1][R#3]	⊥ [R#1][R#3]	⊥ [R#1][R#3]	⊥ [R#1][R#3]	⊥	T	⊥	⊥	T	T
⊥ [R#3]	⊥ [R#3]	⊥ [R#3][R#9]	⊥ [R#3][R#9]	⊥	T	⊥	⊥	T	⊥
⊥ [R#1][R#3]	⊥ [R#1][R#3]	⊥ [R#1][R#3]	⊥ [R#1][R#3]	⊥	T	⊥	⊥	⊥	T
⊥ [R#3]	⊥ [R#3]	⊥ [R#3][R#9]	⊥ [R#3][R#9]	⊥	T	⊥	⊥	⊥	⊥
⊥ [R#4]	T	⊥ [R#8][R#4]	⊥ [R#8]	⊥	⊥	T	T	T	T
⊥ [R#4]	T	⊥ [R#8][R#9][R#4]	⊥ [R#8][R#9]	⊥	⊥	T	T	T	⊥
⊥ [R#1][R#4]	⊥ [R#1]	⊥ [R#1][R#3][R#8][R#4]	⊥ [R#1][R#3][R#8]	⊥	⊥	T	T	⊥	T
⊥ [R#3][R#4]	⊥ [R#3]	⊥ [R#3][R#8][R#9][R#4]	⊥ [R#3][R#8][R#9]	⊥	⊥	T	T	⊥	⊥
T	T	⊥ [R#8]	⊥ [R#8]	⊥	⊥	T	⊥	T	T
T	T	⊥ [R#8][R#9]	⊥ [R#8][R#9]	⊥	⊥	T	⊥	T	⊥
⊥ [R#1]	⊥ [R#1]	⊥ [R#1][R#8]	⊥ [R#1][R#8]	⊥	⊥	T	⊥	⊥	T
T [R#2]	T [R#2]	⊥ [R#8][R#9]	⊥ [R#8][R#9]	⊥	⊥	T	⊥	⊥	⊥
⊥ [R#1][R#4]	⊥ [R#1]	⊥ [R#1][R#8][R#4]	⊥ [R#1][R#8]	⊥	⊥	⊥	T	T	T
⊥ [R#4]	T [R#2]	⊥ [R#8][R#9][R#4]	⊥ [R#8][R#9]	⊥	⊥	⊥	T	T	⊥
⊥ [R#1][R#3][R#4]	⊥ [R#1][R#3]	⊥ [R#1][R#3][R#8][R#4]	⊥ [R#1][R#3][R#8]	⊥	⊥	⊥	T	⊥	T
⊥ [R#3][R#4]	⊥ [R#3]	⊥ [R#3][R#8][R#9][R#4]	⊥ [R#3][R#8][R#9]	⊥	⊥	⊥	T	⊥	⊥
⊥ [R#1]	⊥ [R#1]	⊥ [R#1][R#8]	⊥ [R#1][R#8]	⊥	⊥	⊥	⊥	T	T
T [R#2]	T	⊥ [R#8][R#9]	⊥ [R#8][R#9]	⊥	⊥	⊥	⊥	T	⊥
⊥ [R#1]	⊥ [R#1]	⊥ [R#1][R#8]	⊥ [R#1][R#8]	⊥	⊥	⊥	⊥	⊥	T
T	T	⊥ [R#8][R#9]	⊥ [R#8][R#9]	⊥	⊥	⊥	⊥	⊥	⊥

Table 9. Evaluation of cases with predicate 'Delivery' (3/4)

Desired Result (Reasons) under Access cases at the destination site without the initial master data.	Desired Result (Reasons) under Access cases at the destination site with the initial master data.	Own (u,m)	Dep (m,s_s)	Con (u,s_s)	Dep (m,s_d)	Con (u,s_d)	Dlv (m,s_s,s_d)
⊥ [R#4]	T	T	T	T	T	T	T
⊥ [R#4][R#9]	T [R#9][R#7]	T	T	T	T	T	⊥
⊥ [R#1][R#3][R#4]	⊥ [R#1][R#3]	T	T	T	T	⊥	T
⊥ [R#3][R#4][R#9]	⊥ [R#3][R#9]	T	T	T	T	⊥	⊥
T	T [R#5]	T	T	T	⊥	T	T
⊥ [R#9]	⊥ [R#9]	T	T	T	⊥	T	⊥
⊥ [R#1]	⊥ [R#1]	T	T	T	⊥	⊥	T
⊥ [R#9]	⊥ [R#9]	T	T	T	⊥	⊥	⊥
⊥ [R#1][R#3][R#4]	⊥ [R#1][R#3][R#7]	T	T	⊥	T	T	T
⊥ [R#3][R#4][R#9]	⊥ [R#3][R#9][R#7]	T	T	⊥	T	T	⊥
⊥ [R#1][R#3][R#4]	⊥ [R#1][R#3]	T	T	⊥	T	⊥	T
⊥ [R#3][R#4][R#9]	⊥ [R#3][R#9]	T	T	⊥	T	⊥	⊥
⊥ [R#1][R#3]	⊥ [R#1][R#3]	T	T	⊥	⊥	T	T
⊥ [R#3][R#9]	⊥ [R#3][R#9]	T	T	⊥	⊥	T	⊥
⊥ [R#1][R#3]	⊥ [R#1][R#3]	T	T	⊥	⊥	⊥	T
⊥ [R#3][R#9]	⊥ [R#3][R#9]	T	T	⊥	⊥	⊥	⊥
⊥ [R#8][R#4]	T [R#8][R#7]	T	⊥	T	T	T	T
⊥ [R#8][R#9][R#4]	T [R#8][R#9][R#7]	T	⊥	T	T	T	⊥
⊥ [R#1][R#3][R#8][R#4]	⊥ [R#1][R#3][R#8]	T	⊥	T	T	⊥	T
⊥ [R#3][R#8][R#9][R#4]	⊥ [R#3][R#8][R#9]	T	⊥	T	T	⊥	⊥
⊥ [R#8]	⊥ [R#8]	T	⊥	T	⊥	T	T
⊥ [R#8][R#9]	⊥ [R#8][R#9]	T	⊥	T	⊥	T	⊥
⊥ [R#1][R#8]	⊥ [R#1][R#8]	T	⊥	T	⊥	⊥	T
⊥ [R#8][R#9]	⊥ [R#8][R#9]	T	⊥	T	⊥	⊥	⊥
⊥ [R#1][R#8][R#4]	⊥ [R#1][R#8][R#7]	T	⊥	⊥	T	T	T
⊥ [R#8][R#9][R#4]	T [R#8][R#9][R#7]	T	⊥	⊥	T	T	⊥
⊥ [R#1][R#3][R#8][R#4]	⊥ [R#1][R#3][R#8]	T	⊥	⊥	T	⊥	T
⊥ [R#3][R#8][R#9][R#4]	⊥ [R#3][R#8][R#9]	T	⊥	⊥	T	⊥	⊥
⊥ [R#1][R#8]	⊥ [R#1][R#8]	T	⊥	⊥	⊥	T	T
⊥ [R#8][R#9]	⊥ [R#8][R#9]	T	⊥	⊥	⊥	T	⊥
⊥ [R#1][R#8]	⊥ [R#1][R#8]	T	⊥	⊥	⊥	⊥	T
⊥ [R#8][R#9]	⊥ [R#8][R#9]	T	⊥	⊥	⊥	⊥	⊥

Table 10. Evaluation of cases with predicate 'Delivery' (4/4)

Desired Result (Reasons) under Access cases at the destination site without the initial master data.	Desired Result (Reasons) under Access cases at the destination site with the initial master data.	Own (u,m)	Dep (m,s_s)	Con (u,s_s)	Dep (m,s_d)	Con (u,s_d)	Dlv (m,s_s,s_d)
⊥ [R#4]	T [R#6]	⊥	T	T	T	T	T
⊥ [R#9][R#4]	T [R#9][R#6][R#7]	⊥	T	T	T	T	⊥
⊥ [R#1][R#3][R#4]	⊥ [R#1][R#3]	⊥	T	T	T	⊥	T
⊥ [R#3][R#4][R#9]	⊥ [R#3][R#9]	⊥	T	T	T	⊥	⊥
T [R#6]	T [R#6]	⊥	T	T	⊥	T	T
⊥ [R#9]	⊥ [R#9]	⊥	T	T	⊥	T	⊥
⊥ [R#1]	⊥ [R#1]	⊥	T	T	⊥	⊥	⊥
⊥ [R#9]	⊥ [R#9]	⊥	T	T	⊥	⊥	⊥
⊥ [R#1][R#3][R#4]	⊥ [R#1][R#3][R#6][R#7]	⊥	T	⊥	T	T	T
⊥ [R#3][R#4][R#9]	⊥ [R#3][R#9][R#6][R#7]	⊥	T	⊥	T	T	⊥
⊥ [R#1][R#3][R#4]	⊥ [R#1][R#3]	⊥	T	⊥	T	⊥	T
⊥ [R#3][R#4][R#9]	⊥ [R#3][R#9]	⊥	T	⊥	T	⊥	⊥
⊥ [R#1][R#3]	⊥ [R#1][R#3]	⊥	T	⊥	⊥	T	T
⊥ [R#3][R#9]	⊥ [R#3][R#9]	⊥	T	⊥	⊥	T	⊥
⊥ [R#1][R#3]	⊥ [R#1][R#3]	⊥	T	⊥	⊥	⊥	T
⊥ [R#3][R#9]	⊥ [R#3][R#9]	⊥	T	⊥	⊥	⊥	⊥
⊥ [R#8][R#4]	T [R#8][R#6][R#7]	⊥	⊥	T	T	T	T
⊥ [R#8][R#9][R#4]	T [R#8][R#9][R#6][R#7]	⊥	⊥	T	T	T	⊥
⊥ [R#1][R#3][R#8][R#4]	⊥ [R#1][R#3][R#8]	⊥	⊥	T	T	⊥	T
⊥ [R#3][R#8][R#9][R#4]	⊥ [R#3][R#8][R#9]	⊥	⊥	T	T	⊥	⊥
⊥ [R#8]	⊥ [R#8]	⊥	⊥	T	⊥	T	T
⊥ [R#8][R#9]	⊥ [R#8][R#9]	⊥	⊥	T	⊥	T	⊥
⊥ [R#1][R#8]	⊥ [R#1][R#8]	⊥	⊥	T	⊥	⊥	T
⊥ [R#8][R#9]	⊥ [R#8][R#9]	⊥	⊥	T	⊥	⊥	⊥
⊥ [R#1][R#8][R#4]	⊥ [R#1][R#8][R#6][R#7]	⊥	⊥	⊥	T	T	T
⊥ [R#8][R#9][R#4]	T [R#8][R#9][R#6][R#7]	⊥	⊥	⊥	T	T	⊥
⊥ [R#1][R#3][R#8][R#4]	⊥ [R#1][R#3][R#8]	⊥	⊥	⊥	T	⊥	T
⊥ [R#3][R#8][R#9][R#4]	⊥ [R#3][R#8][R#9]	⊥	⊥	⊥	T	⊥	⊥
⊥ [R#1][R#8]	⊥ [R#1][R#8]	⊥	⊥	⊥	⊥	T	T
⊥ [R#8][R#9]	⊥ [R#8][R#9]	⊥	⊥	⊥	⊥	T	⊥
⊥ [R#1][R#8]	⊥ [R#1][R#8]	⊥	⊥	⊥	⊥	⊥	T
⊥ [R#8][R#9]	⊥ [R#8][R#9]	⊥	⊥	⊥	⊥	⊥	⊥

4   Identify the Set of the Logical Formulas

4.1   Outline

In this section, we derive the set of formulas which can satisfiy the desired results defined in section.3. The simplified constraints are already defined in the previous section 3.2. Thus, in section 4.2, we define the set of formulas in regards to the constraints derived from the combinations of three predicates as in possiblity-3 and 5. On the other hand, in section 4.3 we describe the set of formulas extracted under relationship with the predicate 'Delivery'.

4.2   Set of the Additional Formulas Derived in the General Cases

The previous table.4 shows the desired results of evaluated cases about the co-occurrence among the previous three predicates 'Own','Deploy' and 'Contract'. Accordingly, we can derive the following four formulas. Under the interpretation of 'Allowable', the formulas (20), (21) are identified. (20) means the constraint when the antecedent of the conditional is true, whereas (21) means another constraint for the false case.

$$\vDash_{Allowable} \forall u(u \in \{\mathbf{U}\}), \exists m(m \in \{\mathbf{M_O}\} \cup \{\mathbf{M_C}\}), \exists s(s \in \{\mathbf{S}\})$$
$$\left(Own(u,m) \wedge Dep(m,s) \rightarrow Con(u,s)\right) \quad (20)$$

$$\vDash_{Allowable} \forall u(u \in \{\mathbf{U}\}), \exists m(m \in \{\mathbf{M_O}\} \cup \{\mathbf{M_C}\}), \exists s(s \in \{\mathbf{S}\})$$
$$\left(\neg(Own(u,m) \wedge Dep(m,s)) \rightarrow (\neg Con(u,s) \vee Con(u,s))\right) \quad (21)$$

On the other hand, the formulas (22), (23) are extracted under the interpretation of 'Operational'.

$$\vDash_{Operational} \forall u(u \in \{\mathbf{U}\}), \exists m(m \in \{\mathbf{M_O}\} \cup \{\mathbf{M_C}\}), \exists s(s \in \{\mathbf{S}\})$$
$$\left(Own(u,m) \wedge Dep(m,s) \rightarrow Con(u,s)\right) \quad (22)$$

$$\vDash_{Operational} \forall u(u \in \{\mathbf{U}\}), \exists s(s \in \{\mathbf{S}\})$$
$$\left(\left(Con(u,s) \rightarrow \exists m(m \in \{\mathbf{M_O}\} \cup \{\mathbf{M_C}\})(Own(u,m) \wedge Dep(m,s))\right) \wedge Con(u,s)\right) \quad (23)$$

The previous table.5 shows the desired results of evaluated cases about the co-occurrence among the previous three predicates 'Disclose ','Deploy' and 'Contract'. Accordingly, we can derive the following formulas. Under the interpretation of 'Allowable', the formulas (24) are identified because of a tautology.

$$\vDash_{Allowable} \forall u(u \in \{\mathbf{U}\}), \forall m(m \in \{\mathbf{M_O}\} \cup \{\mathbf{M_C}\}), \forall s(s \in \{\mathbf{S}\})$$
$$\left((Dis(m,u) \vee Dep(m,s) \vee Con(u,s)) \vee \neg(Dis(m,u) \vee Dep(m,s) \vee Con(u,s))\right) \quad (24)$$

On the other hand, the formulas (25), (26) are extracted under the interpretation of 'Operational'.

$$\vDash_{Operational} \forall m(m \in \{\mathbf{M_O}\} \cup \{\mathbf{M_C}\}), \exists u(u \in \{\mathbf{U}\}), \exists s(s \in \{\mathbf{S}\})$$
$$\left(Dis(m,u) \wedge Dep(m,s) \rightarrow Con(u,s)\right) \quad (25)$$

$$\vDash_{Operational} \forall u(u \in \{\mathbf{U}\}), \exists s(s \in \{\mathbf{S}\})$$

$$\Big(\big(Con(u,s) \to \exists m(m \in \{\mathbf{M_O}\} \cup \{\mathbf{M_C}\})(Dis(m,u) \wedge Dep(m,s))\big) \wedge Con(u,s)\Big) \quad (26)$$

4.3 Set of the Formulas Derived in the Cloud Computing

When identifying the set of formulas related to the predicate 'Delivery' illustrated in tables.7, 8, 9 and 10, table.6 could provide the descriptions to identify the set of the elemental formulas. Therefore, at first we will derive these elemental formulas, then, define the whole of the desired formulas linked with the tables.7, 8, 9 and 10 by integrating the previous elemental ones.

Table 11. Evaluation of cases related to [R#1] and [R#2]

$Dlv(m, s_s, s_d)$	$Con(u,s_s) \wedge Con(u,s_d)$	Desired Result	Reasons
T	T	T	Obvious.
T	$\perp$	$\perp$	Obvious as [R#1].
$\perp$	T	T	Obvious.
$\perp$	$\perp$	T	Obvious as [R#2].

Table.11 shows the desired results of evaluated cases about [R#1] and [R#2] in the previous table.6. In these cases, both sites as 'Source' and 'Destination' must be considered when evaluating 'Contract'. The reasons for the desired results in the other cases except the cases of [R#1] and [R#2] are obvious. We need a more thorough consideration about the interpretation which could satisfy the specified desired results in table.11. Accordingly, under our assumption with a more generalization, we could define the formulas expressed as (27) and (28). Furthermore, we tag the formula (27) with the following symbol $r_{11}(m, s_s, s_d, u)$, whereas, the formula (28) as the symbol $r_{12}(m, s_s, s_d, u)$.

$$\vDash \forall m(m \in \{\mathbf{M_O}\} \cup \{\mathbf{M_C}\}), \exists s_s, \exists s_d(s_s, s_d \in \{\mathbf{S}\}, s_s \neq s_d)$$

$$\Big(Dlv(m, s_s, s_d) \to \exists u(Con(u, s_s) \wedge Con(u, s_d))\Big) \quad (27)$$

$$\vDash \forall m(m \in \{\mathbf{M_O}\} \cup \{\mathbf{M_C}\}), \forall s_s, \forall s_d(s_s, s_d \in \{\mathbf{S}\}, s_s \neq s_d)$$

$$\Big(\neg Dlv(m, s_s, s_d) \to \forall u\big(\neg(Con(u, s_s) \wedge Con(u, s_d)) \vee (Con(u, s_s) \wedge Con(u, s_d))\big)\Big) \quad (28)$$

Table 12. Evaluation of cases related to [R#3]

$Dep(m,s_d)$	$Con(u,s_s) \wedge Con(u,s_d)$	Desired Result	Reasons
$\top$	$\top$	$\top$	Obvious.
$\top$	$\bot$	$\bot$	Obvious as [R#3].
$\bot$	$\top$	$\top$	Only contracts Case, Possible.
$\bot$	$\bot$	$\top$	Obvious.

Table.12 shows the desired results of evaluated cases about [R#3] in the previous table.6. In these cases, both sites as 'Source' and 'Destination' must be also considered when evaluating 'Contract'. Here, the case where a contract exists at sites without any deployments is allowable. The reasons for the desired results in the other cases except the above are obvious. We also need a more thorough consideration about the interpretation which could satisfy the specified desired results in table.12. Accordingly, under our assumption with a more generalization, we could define the formulas expressed as (29) and (30). Furthermore, we also tag the formula (29) with the following symbol $r_{31}(m, s_s, s_d, u)$, whereas, the formula (30) as the symbol $r_{32}(m, s_s, s_d, u)$.

$$\vDash \forall m(m \in \{\mathbf{M_O}\} \cup \{\mathbf{M_C}\}), \exists s_s, \exists s_d (s_s, s_d \in \{\mathbf{S}\}, s_s \neq s_d) \\ \left(Dep(m, s_d) \rightarrow \exists u \left(Con(u, s_s) \wedge Con(u, s_d) \right) \right) \quad (29)$$

$$\vDash \forall m(m \in \{\mathbf{M_O}\} \cup \{\mathbf{M_C}\}), \forall s_s, \forall s_d (s_s, s_d \in \{\mathbf{S}\}, s_s \neq s_d) \\ \left(\neg Dep(m, s_d) \rightarrow \forall u \left(\neg \left(Con(u, s_s) \wedge Con(u, s_d) \right) \vee \left(Con(u, s_s) \wedge Con(u, s_d) \right) \right) \right) \quad (30)$$

Table 13. Evaluation of cases related to [R#4]

$Dep(m,s_s)$	$Dep(m,s_d)$	Desired Result	Reasons
$\top$	$\top$	$\bot$	Obvious as [R#4].
$\top$	$\bot$	$\top$	Obvious.
$\bot$	$\top$	$\bot$	Obvious as [R#4].
$\bot$	$\bot$	$\top$	Don't Care.

Table.13 shows the desired results of evaluated cases about [R#4] in the previous table.6. In these cases, both sites as 'Source' and 'Destination' must be also considered when evaluating 'Deployments'. Here, the case where the destination site already has a replica of the specified master data, the desired result must be false. And we will regard the case where the result should originally be 'Don't Care', as true. We also need a more thorough consideration about the interpretation which could satisfy the specified desired results in table.13. Accordingly, under our assumption with a more generalization, we could define the formula expressed as (31). Furthermore, we also tag the formula (31) with the following symbol$r_{41}(m, s_s, s_d)$.

$$\vDash \forall m(m \in \{\mathbf{M_O}\} \cup \{\mathbf{M_C}\}), \forall s_s, \exists s_d (s_s, s_d \in \{\mathbf{S}\}, s_s \neq s_d)$$
$$\left(Dep(m, s_d) \rightarrow \left(\neg Dep(m, s_s) \wedge Dep(m, s_s)\right)\right) \quad (31)$$

Table 14. Evaluation of cases related to [R#7]

C : Any Formula	Dep(m,s_d) ∧ Con(u,s_d)	Desired Result	Reasons
T	T	T	Obvious as [R#7].
T	⊥	T	Need to maintain the original state.
⊥	T	T	Need to change for [R#7].
⊥	⊥	⊥	Need to maintain the original state.

As for [R#5] and [R#6] in the previous table.6, we omitted them because they are non-descriptive to express as formulas. In particular, [R#6] has no corresponding class explicitly in Fig.4. Table.14 shows the desired results of evaluated cases about [R#7] in the previous table.6. In the case where the destination site already has a replica of the specified master data, the desired result would be set as true, whereas in the other cases, the original value of the opposite formula would be maintained. We also need a more thorough consideration about the interpretation which could satisfy the specified desired results in table.14. Accordingly, under our assumption with a more generalization, we could define the formula expressed as (32). Here, we assume that the symbol $\{\mathbf{C'}\}$ expresses a set of any formulas except ones belonging to the union set$\{\mathbf{C}\} \cup \{\mathbf{C_{cloud}}\}$.

$$\vDash \forall c(c \in \{\mathbf{C'}\} \cup \{\mathbf{C}\} \cup \{\mathbf{C_{cloud}}\}), \forall m(m \in \{\mathbf{M_O}\} \cup \{\mathbf{M_C}\}), \forall s_d(s_d \in \{\mathbf{S}\}), \forall u(u \in \{\mathbf{U}\})$$
$$\left(c \vee (Dep(m, s_d) \wedge Con(u, s_d))\right) \quad (32)$$

Table 15. Evaluation of cases related to [R#8]

C : Any Formula	$Dep(m,s_s)$	Desired Result	Reasons
T	T	T	Need to maintain the original state.
T	$\perp$	$\perp$	Obvious as [R#8].
$\perp$	T	$\perp$	Need to maintain the original state.
$\perp$	$\perp$	$\perp$	Obvious as [R#8].

Table.15 shows the desired results of evaluated cases about [R#8] in the previous table.6. In the case where there are no specified master data at the source site, the desired result would be set as false, whereas in the other cases, the original value of the opposite formula would be maintained. We also need a more thorough consideration about the interpretation which could satisfy the specified desired results in table.15. Accordingly, under our assumption with a more generalization, we could define the formula expressed as (33).

$$\vDash \forall c(c \in \{\mathbf{C'}\} \cup \{\mathbf{C}\} \cup \{\mathbf{C_{cloud}}\}), \forall m(m \in \{\mathbf{M_O}\} \cup \{\mathbf{M_C}\}), \forall s_s(s_s \in \{\mathbf{S}\})$$
$$\left(c \wedge Dep(m, s_s)\right) \quad (33)$$

Table 16. Evaluation of cases related to [R#9]

C : Any Formula	$Dlv(m,s_s,s_d)$	Desired Result	Reasons
T	T	T	Need to maintain the original state.
T	$\perp$	$\perp$	Obvious as [R#9].
$\perp$	T	$\perp$	Need to maintain the original state.
$\perp$	$\perp$	$\perp$	Obvious as [R#9].

Table.16 shows the desired results of evaluated cases about [R#9] in the previous table.6. In the case where the predicate 'Delivery' is set as false, desired result of the entire formula should also be set in false, whereas in the other cases, the original value of the opposite formula would be maintained. We also need a more thorough consideration about the interpretation which could satisfy the specified desired results in table.16. Accordingly, under our assumption with a more generalization, we could define the formula expressed as (34).

$$\vDash \forall c(c \in \{\mathbf{C'}\} \cup \{\mathbf{C}\} \cup \{\mathbf{C_{cloud}}\}), \forall m(m \in \{\mathbf{M_O}\} \cup \{\mathbf{M_C}\}), \forall s_s, \exists s_d(s_s, s_d \in \{\mathbf{S}\}, s_s \neq s_d)$$
$$\left(c \wedge Dlv(m, s_s, s_d)\right) \quad (34)$$

Based on the multiple formulas from (27) to (34), we could define the entire set of the formulas related to the predicate 'Delivery' illustrated in tables.7, 8, 9 and 10 as mention later. However, the descriptions in regards to quantifiers and domain definitions are omitted here in order to simplify these definitions. The formula expressed as (35) must be satisfied when a pre-deployed master data is banned at the destination site under the interpretation of 'Allowable'.

$$\vDash_{Allowable} (r_{11}(m, s_s, s_d, u) \lor r_{12}(m, s_s, s_d, u)) \land$$
$$(r_{31}(m, s_s, s_d, u) \lor r_{32}(m, s_s, s_d, u)) \land r_{41}(m, s_s, s_d) \quad (35)$$

Conversely, the formula expressed as (36) should be satisfied when a pre-deployed master data at the destination site is accepted under the interpretation of 'Allowable'. This is caused because the tagged elemental formula $r_{41}(m, s_s, s_d)$ is not required.

$$\vDash_{Allowable} (r_{11}(m, s_s, s_d, u) \lor r_{12}(m, s_s, s_d, u)) \land (r_{31}(m, s_s, s_d, u) \lor r_{32}(m, s_s, s_d, u)) \quad (36)$$

Under the interpretation of 'Operational', the formula expressed as (37) must be satisfied instead of previous (35), when a pre-deployed master data is banned at the destination site. This is caused because of stricter constraints.

$$\vDash_{\{Available, Operational\}} \left(\begin{array}{c} (r_{11}(m, s_s, s_d, u) \lor r_{12}(m, s_s, s_d, u)) \land \\ (r_{31}(m, s_s, s_d, u) \lor r_{32}(m, s_s, s_d, u)) \land r_{41}(m, s_s, s_d) \end{array} \right)$$
$$\land Dep(m, s_s) \land Dlv(m, s_s, s_d) \quad (37)$$

Conversely, under the interpretation of 'Operational', the formula expressed as (38) should be satisfied when a pre-deployed master data at the destination site is accepted. This is caused due to the same reason as (36). The tagged elemental formula $r_{41}(m, s_s, s_d)$ is also not required.

$$\vDash_{Operational} \left(\begin{array}{c} (r_{11}(m, s_s, s_d, u) \lor r_{12}(m, s_s, s_d, u)) \land \\ (r_{31}(m, s_s, s_d, u) \lor r_{32}(m, s_s, s_d, u)) \end{array} \right) \land Dep(m, s_s) \land Dlv(m, s_s, s_d) \quad (38)$$

Finally, under the interpretation of 'Available', this originally arises only in the case where a pre-deployed master data at the destination site is accepted. Therefore, the formula expressed as (39) must be satisfied. On the other hand, under the condition where a pre-deployed master data at the destination site is banned, (37) should be adopted because of being the same situation under the interpretation of 'Operational'.

$$\vDash_{Available} \left(\begin{array}{c} ((r_{11}(m, s_s, s_d, u) \lor r_{12}(m, s_s, s_d, u)) \land (r_{31}(m, s_s, s_d, u) \lor r_{32}(m, s_s, s_d, u))) \\ \land Dep(m, s_s) \land Dlv(m, s_s, s_d) \end{array} \right)$$
$$\lor (Dep(m, s_d) \land Con(u, s_d)) \quad (39)$$

5   Points to Consider

In this paper, we defined the multiple interpretations and the set of formulas expressing the inductive semantic model of constraints as an application of the first order predicate logic. However, that only remains at defining a semantic model as Fig.3 shows. We need to verify them and prove the consistency and completeness by the first order predicate calculus after defining the syntax with the set of the logical axioms and inference rules. During that procedure, the set of tautologies will be derived. When encountering any failures in the derivation, there are potentially two major suspected causes. The first cause is to fail to identify all of the necessary formulas which are more than the specified set of formulas used here. The second cause depends on incomplete definitions of the predicates and classes. For instance, the related classes around authorization are not explicitly depicted in Fig.4. This is not reflected because we need more classes such as 'Object', 'Operation', 'Role', and 'Permission' and so on. They are mandatory when we try to import the class set of the authorization, especially Role Based Access Control [5]. However, as a fact there was an ambiguity in the entire constraints which was our original concern to begin with in the scope defined in Fig. 4 even with the lack of discussion about authorization. Thus, we can expect that these ambiguities and the absence of logical completeness could be solved during the previous procedure for revealing the set of tautologies.

We will briefly touch on related works due to limited space. The most related study might be one carried out by M.Comerio et al [6]. As for this, they have analyzed the requirements of Data Quality Engineering (DQE) when applying this DQE into the Cloud environments. Additionally they propose their conceptual architecture to realize it. The prominent contributions in this study are to survey our targeted area comprehensively and to define the trial concepts at the initial phase, in particular on the elemental functionalities based on the general process appearing in DQE. However, they do not focus only on MDM as our target. Furthermore, they do not explicitly deal with the constraints matter caused by security requirements. Therefore, they stand on another point from ours.

6   Conclusions

In this paper, we have presented the precise constraints model by applying the first order predicate logic and inductive approach. Further we give explainable description of its contexts to clarify the semantics. In consideration of future work, we need to verify the syntax after adding the set of logical axioms to prove the consistency and completeness. And, also we need to import other classes such as authorization to make sure there is the harmonization with the existing rules.

Finally, we mention the potential usage of this semantic model after achieving the completeness. This model is expected to be contributable in defining an implementable language for operational constraints. As our model does not deal with the definitions around the authorization explicitly, we could expect this might be mapped as an upper function beyond the existing secure functionality. However, as this model relies on the capability of first order language, all of the mapping or linked targets must be enumerated as an interpretation. When considering the actual operational and machinery environments for implementation, issues will still be caused as to how we should protect an instance of the description along this model and how we should guarantee the security. Furthermore, it is sufficiently considerable that a cleansing process would become a complicated operation including the intricate workflows even though we have already just touched on the cleansing. It is still vague whether our model potentially has the descriptive capability to handle these requirements. Therefore, it is obviously natural to execute both the following items in parallel; the first is to verify the descriptive capability of our model by deriving the previous set of tautologies. The second is to extend the scope of modeled classes and import them in order to enhance the descriptive capability more.

References

1. Mosley, M., Brackett, M., Earley, S., Henderson, D.: The DAMA Guide to The Data Management Body of Knowledge (DAMA-DMBOK Guide): The Data Management association (2010)
2. Dreibelbis, A., Hechler, E., et al.: Enterprise Master Data Management. IBM Press (2008)
3. Kikuchi.S.: Structural Analysis of Issues in Exchanging Qualified Master Data under Cloud Computing. Submitted to International Journal of Computational Science and Engineering, 1–13 (December 5, 2012) (to appear)
4. Stoitsev, V., Grefen, P.: Business Process Technology and the Cloud: defining a Business Process Cloud Platform: BETA Working Paper 2012 (2012), http://cms.ieis.tue.nl/Beta/Files/WorkingPapers/wp_393.pdf
5. American National Standard for Information Technology – Role Based Access Control, INCITS 359-2004 (2004)
6. Comerio, M., Trong, H.L., Batini, C., Dustdar, S.: Service-oriented Data Quality Engineering and Data Publishing in the Cloud. In: 2010 IEEE International Conference on Service-Oriented Computing and Application, SOC 2010 (2010)

Efficient Selection of Various k-Objects for a Keyword Query Based on MapReduce Skyline Algorithm

Md. Anisuzzaman Siddique and Yasuhiko Morimoto

Hiroshima University,
1-7-1 Kagamiyama, Higashi-Hiroshima, 739-8521, Japan
anis_cst@yahoo.com,
morimoto@mis.hiroshima-u.ac.jp
http://www.morimo.com/morimo-ken

Abstract. Recently, keyword-based query interface is a de facto standard for information retrieval. A user gives a keyword and gets necessary objects that are closely related to the keyword. How to select the necessary objects is one of the most important problem in database literature. Top-k query is popular method to select important objects from large candidate objects. A user specfies a scoring function and k. Then, the top-k query selects the k objects based on the scoring function. However, each user may have different scoring function to select the top-k object, which means the top-k objects are valuable only for users who share the same scoring function. In this paper, we propose k-objects selection function that selects various k objects that are preferable for all users who may have different scoring function. We applied the idea of skyline queries to select the k objects in this paper. We also considered efficient computation by using MapReduce flamework.

Keywords: Skyline Query, Top-k Query, MapReduce, Mobile Phone Interface.

1 Introduction

Recent computing infrastructure makes a large amount of information available to consumers, which is creating information overload problems. Many kinds of information retrieval functions have been considered to handle the problems. Among them *top-k* queries [1], [2], [3], [4] have been widely used to select preferable objects from large data source.

A user specifies a scoring function and k. Then, top-k query selects the k objects based on the scoring function. After defining a scoring function, a user does not have to specify complicated query conditions to retrieve necessary objects. Therefore, top-k query is suitable query interface for mobile phones. However, each user may have different scoring function to select the top-k object, which means the top-k objects are valuable only for users who share the same scoring function.

A. Madaan, S. Kikuchi, and S. Bhalla (Eds.): DNIS 2014, LNCS 8381, pp. 40–52, 2014.

On the other hand, *skyline query* [11], [12], [9], [15], which is also a popular information retrieval function, have been used for eliminating dominated objects. Skyline can select objects that are preferable for all users who have different scoring function. However, it may retrieve too many or too small objects.

In this paper, we propose k-objects selection function that selects various k objects that are preferable for all users who may have different scoring function. Recently, we often have to retrieve necessary objects using a mobile phone. In such situation, it is difficult to specify complicated query condition and we want to retrieve objects by specifying a keyword and the number of objects, say k. The proposed function must be useful for such situation.

To clarify our proposal, consider following example that illustrates top-k query and skyline query.

Example (Top-k and Skyline queries)

Consider a hotel database DB that has "price" and "distance" attributes as in Figure 1. To reduce the data to the top-1 hotel, one user may evaluate hotels with his/her own scoring function, e.g., with the lowest aggregate of price and distance, which we call "scoring function 1" as in Figure 1. Alternatively, another user may look for the shortest distance hotel and do not care about price, which we call "scoring function 2". Although these scoring function clearly determine top-1 hotel, it is difficult for a mobile phone user to define such an exact scoring function.

Meanwhile, observe that for both users, hotel O_5 is a better choice than O_4, as O_5 is better or equals to O_4 in all attributes, i.e., O_5 dominate O_4. Based on this intuition, we can identify "skyline objects" that are not dominated by any other objects. A user who wants to choose a hotel can consider the hotels in the skyline $\{O_1, O_3, O_5, O_7\}$, since the skyline contains the best hotel of any linear scoring functions.

As illustrated in the above example, the skyline queries do not require a scoring function F and simply find a common subset of non-dominated objects for all linear scoring functions. This intuitive nature of query formulation has been a key strength of skyline queries, compared to top-k queries that requires users to formulate F. On the other hand, we cannot control the number of retrieved objects. Skyline queries may retrieve too many objects especially in high dimensional data sets. It also may retrieve too small objects if there is a very strong object in a database.

It is non-trivial to identify truly interesting objects from large skyline object set. We combine the strength of both skyline and top-k query and propose an algorithm to select various k objects so that the k objects contain the top-k' ($k' \ll k$) objects for all users.

We consider an efficient algorithm to select the k objects with MapReduce framework. MapReduce framework is scalable, fault tolerant, cost effective and easy to use. Non-experts can deploy their applications simply by implementing a

Map function and a Reduce function, whereas the data management, task management, fault tolerance and recovery are provided transparently. The success of MapReduce has reported in many applications [21], [22], [23].

In this paper, we propose an efficient parallel algorithm, which selects various k objects using MapReduce.

ID	Price (rank)	Distance (rank)
O_1	3 (1)	17 (6)
O_2	17 (9)	19 (8)
O_3	9 (4)	7 (2)
O_4	16 (8)	11 (4)
O_5	13 (6)	3 (1)
O_6	10 (5)	7 (2)
O_7	8 (3)	12 (5)
O_8	13 (6)	18 (7)
O_9	3 (1)	20 (9)

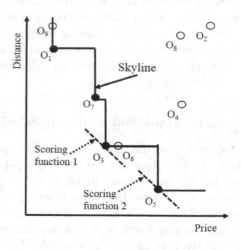

a) Symbolic Dataset   b) Skyline

Fig. 1. Skyline Example

The rest of this paper is organized as follows: Section 2 reviews related work. Section 3 presents the notions and properties of top-k skyline objects computation. We provide detailed examples and analysis of our algorithms in Section 4. We experimentally evaluate the proposed algorithms in Section 5 under a variety of settings. Finally, Section 6 concludes the paper.

2 Related Work

We summarize the related works from three aspects: skyline query processing, top-k query processing, and MapReduce based query processing.

2.1 Skyline Query Processing

Skyline queries have been studied since 1960s in the theory field where skyline objects are known as Pareto sets and admissible objects [7] or maximal vectors [5]. However, earlier algorithms such as [5], [6] are inefficient when there are many data objects in a high dimensional space.

Borzsonyi, *et al.* first introduce the skyline operator over large datasets and proposed three algorithms: Block-Nested-Loops (BNL), Divide-and-Conquer (D&C), and B-tree-based schemes [11]. BNL compares each object of the dataset with every other object, and reports it as a result only if any other object does not dominate it. A window W is allocated in main memory, and the input relation is sequentially scanned. In this way, a block of skyline objects is produced in every iteration. In case the window saturates, a temporary file is used to store objects that cannot be placed in W. This file is used as the input to the next pass. *D&C* divides the dataset into several partitions such that each partition can fit into memory. Skyline objects for each individual partition are then computed by a main-memory skyline algorithm. The final skyline is obtained by merging the skyline objects for each partition. Chomicki, *et al.* improved BNL by presorting, they proposed *Sort-Filter-Skyline(SFS)* as a variant of BNL [13]. SFS requires the dataset to be pre-sorted according to some monotone scoring function. Since the order of the objects can guarantee that no object can dominate objects before it in the order, the comparisons of objects are simplified.

Among index-based methods, Tan, *et al.* proposed two progressive skyline computing methods Bitmap and Index [14]. The current most efficient method is *Branch-and-Bound Skyline(BBS)*, proposed by Papadias, *et al.*, which is a progressive algorithm based on the best-first nearest neighbor (BF-NN) algorithm [9]. Instead of searching for nearest neighbor repeatedly, it directly prunes using the R*-tree structure.

Recently, more aspects of skyline computation have been explored. Vlachou, *et al.* introduce the concept of extended skyline set, which contains all data elements that are necessary to answer a skyline query in any arbitrary subspace [15]. Fotiadou, *et al.* mention about the efficient computation of extended skylines using bitmaps in [16]. Chan, *et al.* introduce the concept of *skyline frequency* to facilitate skyline retrieval in high-dimensional spaces [17]. Tao, *et al.* discuss skyline queries in arbitrary subspaces [18].

2.2 Top-k Query Processing

Top-k queries about skyline were studied in [8], [9], [10]. Papadias, *et al.* discussed ranked skyline and k-dominating queries [9]. Given a set of objects in d-dimensional space, ranked skyline specifies a monotone ranking function, and returns k objects in the d-dimensional space which have the smallest (or greatest) scores according to an input function. Given a set of objects in d-dimensional space, k-dominating queries retrieve k objects that dominate the greatest number of objects.

Lin, *et al.* studied representative skyline queries [8]. The problem is to select k objects among all skyline objects according to a pre-defined objective function. The k objects in the output are said to be representative. Representative skyline query finds a set of k objects among all skyline objects such that the number of objects dominated by this set is maximized.

Another definition of representative skyline queries was proposed by Tao, *et al.* [10]. In [10], representative skyline queries is to find k objects (or k representative objects) among all skyline objects such that the sum of the distances between each skyline object and its "closest" representative object is minimized.

All of the above studies are to find k objects given a DB where attribute preferences in the dataset are given. This paper has the following differences. Firstly, we want to find k skyline objects where the attribute preference is not given. Secondly, the benefit of MapReduce (parallel computing) is considered in this paper but not in the above studies.

2.3 MapReduce Based Query Processing

Computing the skyline or its variants is challenging today since there is an increasing trend of applications expected to deal with *big data*. For such data intensive applications, the MapReduce [21], [22], [23] framework has recently attracted a lot of attention. MapReduce is a programming model that allows easy development of scalable parallel applications to process big data on large clusters of commodity machines. Ideally, a MapReduce system should achieve a high degree of load balancing among the participating machines, and minimize the space uses, CPU and I/O time, and network transfer at each machine. However, there exist some recent works on skyline computation using MapReduce [20], [19]. All of these works focus on efficient computation but did not give any idea how to choose k objects from skyline.

In MapReduce framework, the implementation of Mappers and Reducers are completely independent of each other without communication among Mappers or Reducers. When processing this type of applications on MapReduce, there are large amount unpromising data intermediate data to be transferred. These unpromising intermediate data will be finally abandoned anyhow, leading to the waste of disk access, network bandwidth and CPU resources. To filter unpromising data and reduce the amount of intermediate data and the waste of time processing the unpromising data can be mitigated by applying skyline query search first. Then find the k object from the skyline result using MapReduce framework.

3 k-Objects Selection Problem

3.1   Preliminaries

Given a data space D that is defined by a set of m-attributes $\{a_1, a_2, \cdots, a_m\}$ and there is a database DB on D. Let $\{O_1, O_2, \cdots, O_n\}$ be n objects of DB. Without loss of generality, we assume that each attribute has non-negative numerical values. We also assume smaller value is preferable in each attribute. We use $O_i.a_j$ to denote the j-th attribute's value of O_i.

For objects O_i and O_j, an object O_i is said to *dominate* another object O_j with respect to D, denoted by $O_i \leq O_j$, if $O_i.a_s \leq O_j.a_s$ for all attributes

$(s = 1, \cdots, n)$ and $O_i.a_t < O_j.a_t$ for at least one attribute $(1 \leq t \leq n)$. We call such O_i as *dominant object* and such O_j as *dominated object* between O_i and O_j. If O_i dominate O_j, then O_i is more preferable than O_j.

Definition *Skyline*: An object $O \in DB$ is in skyline of DB (i.e., a skyline object in DB) if O is not dominated by any other object in DB. The skyline of DB, denoted by $Sky(DB)$, is the set of skyline objects in DB.

For example in the table of Figure 1 where $DB = \{O_1, \cdots, O_9\}$, since O_1, O_3, O_5, and O_7 are not dominated by any object in DB, $Sky(DB)$ is $\{O_1, O_3, O_5, O_7\}$.

The number of objects in a skyline is different for each database. Sometime, the number of objects may be too small for analysis. In such situation, *extended skyline* is can be used to increase result set. For objects O_i and O_j, an object O_i is said to *extended-dominate* another object O_j with respect to D, denoted by $O_i < O_j$, if $O_i.a_s < O_j.a_s$ for all attributes $(s = 1, \cdots, n)$. We call such O_i as *extended-dominant* object and such O_j as *extended-dominated* object between O_i and O_j.

Definition *Extended Skyline*: An object $O \in DB$ is in extended skyline of DB (i.e., a extended skyline object in DB) if O is not extended dominated by any other object in DB. The extended skyline of DB, denoted by $ext\text{-}Sky(DB)$, is the set of extended skyline objects in DB.

In the table in Figure 1, since O_1, O_3, O_5, O_6, O_7, and O_9 are not extended dominated by any objects in DB, $ext\text{-}Sky(DB)$ is equal to $\{O_1, O_3, O_5, O_6, O_7, O_9\}$.

Top-k queries can be defined by a scoring function F, which enables the ranking (ordering) of the data objects. The most important and commonly used scoring function is the ranked linear sum function. Each attribute a_i has an associated rank r_i indicating a_i's relative importance for the query. The aggregated score $F_{\mathbf{w}}(O)$ for object O is defined as a weighted sum of the individual ranks: $F_{\mathbf{w}}(O) = \sum_{i=1}^{m} w_i * r_i$, where $\mathbf{w} = (w_1, ..., w_m)$ is a user given weighting vector.

The result of a top-k query is a list of the k objects that have the top-k ranking values of $F_{\mathbf{w}}$. Consider for the example in Figure 1. By assigning a high weight to attribute "price", we can select cheaper objects. If a user specifies that $k = 2$ and $\mathbf{w} = \{0.9, 0.1\}$. The top-2 result is $\{O_1, O_9\}$. On the other hand, if we assign a high weight to "distance" and specify $\mathbf{w} = \{0.1, 0.9\}$, the top-2 result is $\{O_5, O_3\}$.

Definition *Top-k Query*: Given a positive integer k and a weighting vector $\mathbf{w}$, the result set $TOP_k(\mathbf{w})$ of the top-k query is a set of objects such that $TOP_k(\mathbf{w}) \in DB$, $|TOP_k(\mathbf{w})| = k$ and $\forall O_i, O_j$: $O_i \in TOP_k(\mathbf{w})$, $O_j \in DB - TOP_k(\mathbf{w})$ it holds that $F_{\mathbf{w}}(O_i) \leq F_{\mathbf{w}}(O_j)$.

We can specify the number of retrieved objects by using top-k query. However, to specify a weighting vector is not an easy procedure for a user especially in mobile environment.

Recently, keyword-based query interface is a de facto standard for information retrieval. A user gives a keyword and gets necessary objects that are closely related to the keyword. We assume that a user can specify the number of necessary object k but cannot specify a weighting vector.

3.2 k-Objects Selection

In this section we will explain how to select k-objects from $Sky(DB)$. User needs to specify the number of retrieved objects i.e., k. Then the following two cases can occur.

(Case 1: $|Sky(DB)| \geq k$) The number of skyline objects is larger than user specified number. Then proposed technique use an aggregates function to compute combined rank of $Sky(DB)$ from individual attributes rank (shown in Figure 1). Retrieves k most influential objects as result set. Here we sort $Sky(DB)$ objects according to the ascending order of combined rank and select k object as result.

(Case 2: $|Sky(DB)| < k$) The number of skyline objects is smaller than user specified number. In this case proposed technique use $extSky(DB)$ and calculate combined rank of $extSky(DB)$ by following above procedure. Finally, retrieves k $extSky(DB)$ objects as result set.

4 k-Objects Selection Using MapReduce

The parallel computation of the k-objects in a given data set DB consists of the following three phases. The algorithm is shown in Algorithm 1.

Skyline and Splitting phase: To filter out non-skyline objects effectively, we first compute $Sky(DB)$. Then, we partition the $Sky(DB)$ vertically and store the partitioned skyline objects for further processing.

Map and Ranking phase: In this phase, each Map function generates $(Key, Value)$ pairs, where Key is the object ID and $Value$ is the rank of corresponding attribute value. The output of this phase is $(ID, Rank)$ pairs for each object.

Reduce and Selection phase: A combiner collect all $(ID, Rank)$ pairs for each object to reduce data access between Map and Reducer. The reduce function is invoked in this stage and produces $(Key, Value)$ pairs that are the resultant of the MapReduce job. In this work, each reduce function produces $(ID, list(Rank))$ pairs. Finally, selects k objects from $Sky(DB)$ which have lowest ranks.

4.1   Skyline and Splitting Phase

The implementation of Mappers and Reducers are completely independent in MapReduce flamework. To compute the k-objects selection function on MapReduce, there exist large amount unpromising intermediate data to be transferred. These intermediate data cause bad effect on disk access, network bandwidth, and CPU resources. To mitigate unpromising data (i.e., non-skyline object) effectively, we apply skyline query on DB and compute $Sky(DB)$. In this work, we apply SFS [13] algorithm to compute $Sky(DB)$ efficiently. We choose non index based skyline computation algorithm because index based skyline computation algorithms are not suitable for MapReduce framework. $Sky(DB)$ is shown in Table 1.

Table 1. $Sky(DB)$ Dataset

ID	Price	Distance
O_1	3	17
O_3	9	7
O_5	13	3
O_7	8	12

Since MapReduce is a parallel programming model, the next consideration is how to split the computation job. A straightforward way is to divide the $Sky(DB)$ dataset into several subsets, i.e, horizontal splitting/partitioning. Another way is to split $Sky(DB)$ dataset in a vertical fashion, i.e., each subset containing one single attribute of the dataset. We followed a vertical partitioning strategies in our k-objects selection. For convenience, m-dimensional dataset splitted into m partitions, say $\{S_1, S_2, \cdots, S_m\}$. In order to compute the k-objects, we introduce an additional partition S_{m+1}. Partition S_{m+1} consists the aggregate values of all dimensions for each data object. The necessity of this partition is explained in next section. In our running example, $Sky(DB)$ has two attributes "price" and "distance". We split this $Sky(DB)$ into two partitions called S_1, S_2. And, we prepare another partition S_3, in which sum of "price + distance" for each object are stored. Here, we have three partitions, which means we need at least three Mappers.

4.2 Map and Ranking Phase

In this phase, each Map task independently operates a non-overlapping split of the input file and calls the user-defined Map function to emit a list of Key-$Value$ pairs (K, V) from its local storage in parallel. While Key is usually numeric, the value V can contain arbitrary information. In our algorithm, each Map function produces $(ID, Rank)$ pair according to the sorted value of each attribute. To calculate the rank value, each Map function performs the following procedure. For each S_l ($l = 1, \cdots, m + 1$), sort its attributes in ascending order of their value, then replace the values by their corresponding rank value. Table 2 shows the output of Mappers of our algorithm.

Table 2. $(ID, Rank)$ pairs

$S_1, price$	$S_2, distance$	$S_3, price + distance$
(ID, Rank)	(ID, Rank)	(ID, Rank)
$O_1, 1$	$O_5, 1$	$O_3, 1$
$O_7, 2$	$O_3, 2$	$O_5, 1$
$O_3, 3$	$O_7, 3$	$O_1, 3$
$O_5, 4$	$O_1, 4$	$O_7, 3$

4.3   Reduce and Selection Phase

The *ID-Rank* pairs emitted by all map functions are, then, grouped by *ID*. Table 3 shows the output of the combiner function, which are grouped by *ID*.

Table 3. Group by corresponding Object *ID*

(ID, Rank)	(ID, Rank)
$O_1, 1$	$O_5, 4$
$O_1, 4$	$O_5, 1$
$O_1, 3$	$O_5, 1$
$O_3, 3$	$O_7, 2$
$O_3, 2$	$O_7, 3$
$O_3, 1$	$O_7, 3$

Next, the reduce function is invoked with each distinct *ID* and the list of all values sharing the *ID*, and output *ID-Rank* pairs. It aggregates all the rank values according to distinct object *ID* and produces $(ID, sum(Rank))$ pairs. From Table 3 we see that the Reduce function of our algorithm will produce four $(ID, sum(Rank))$ pairs. They are $(O_1, 8)$, $(O_3, 6)$, $(O_5, 6)$, and $(O_7, 8)$.

Besides the Map and Reduce functions, there exist another function called main function. Main function is responsible to coordinate all of the works of Mappers and Reducers. If an user specifies $k = 2$, then the main function can retrieves the top-2 objects according to the sum value, which are O_3 and O_5.

4.4   Extended k-Objects

Sometimes, the number of skyline objects is smaller than the user specified k. In such situation, extended skyline can be used As we mentioned above, skyline query retrieves $Sky(DB) = \{O_1, O_3, O_5, O_7\}$, while extended skyline query retrieves $ext\text{-}Sky(DB) = \{O_1, O_3, O_5, O_6, O_7, O_9\}$. Then, we select k objects from $ext\text{-}Sky(DB)$ by Algorithm 4.3.

5   Performance Evaluation

We conduct a series of experiments with different data distributions, dimensionalities, data cardinalities and query cardinalities to evaluate the effectiveness and efficiency of our proposed methods. Each experiment is repeated five times and the average result is considered for performance evaluation. Two data distributions are considered as follows:

Anti-correlated: An anti-correlated dataset represents an environment in which, if an object has a small coordinate on some dimension, it tends to have a large coordinate on at least another dimension.

Algorithm 1: k-Objects Selection

Input: Dataset DB; retrieval objects size k.
Output: Selected k-Objects in DB.
Skyline Computation Procedure
  From input dataset DB compute skyline using SFS algorithm.
  Return skyline dataset $Sky(DB)$
Split Procedure
  Split dataset $Sky(DB)$ into (m) partitions $\{S_1, S_2, \cdots, S_m\}$
  Create another partition S_{m+1} from the aggregate values of all dimensions.
Map Procedure
  For each S_l $(l = 1, \cdots, (m+1))$ do
    Sort S_l attribute values in ascending order
    Assign corresponding rank r
    Return $(ID, Rank)$ pairs
Combine Procedure
  Collect $(ID, Rank)$ pairs
  For each $Sky(DB)$ object ID grouped $(ID, Rank)$ pairs
    Return grouped $(ID, Rank)$ pairs
Reduce Procedure
  Collect grouped $(ID, Rank)$ pairs
  For each grouped $(ID, Rank)$ pair produce $(ID, sum(Rank))$ pairs
    Return $(ID, sum(Rank))$ pairs
Finally, *main function* outputs desired k objects according to ascending $sum(Rank)$

Independent: For this type of dataset, all attribute values are generated independently using uniform distribution. Under this distribution, the total number of non-dominating objects is between that of the correlated and the anti-correlated datasets.

Effect of Dimensionality. We first study the effect of dimensionality on our techniques. We fix the data cardinality to 100k and vary dataset dimensionality n ranges from 2 to 8 and set query number to 0.5k. The runtime results for this experiment are shown in Figure 2(a) and (b).

Effect of Cardinality. For this experiment, we fix the data domensionality to 6 and vary dataset cardinality ranges from 100k to 500k and set query number to 0.5k. Figure 3(a) and (b) shows the performance on anti-correlated and independent datasets. The proposed techniques are highly affected by data cardinality. If the data cardinality increases then the performances decreases.

Effect of Query Number. For this experiment, we use both types ddata distributions with varying query number ranges from 0.5k to 2k and set the value of n to 6 and data cardinality to 100k. Figure 4(a) and (b) shows the time to answer all queries. The result shows that if the query number increases the computation time of proposed methods also increases.

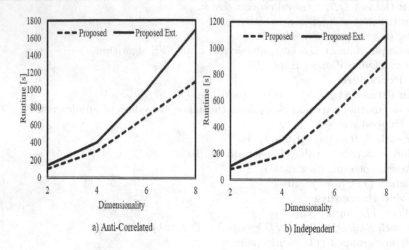

a) Anti-Correlated

b) Independent

Fig. 2. Performance for different data dimension

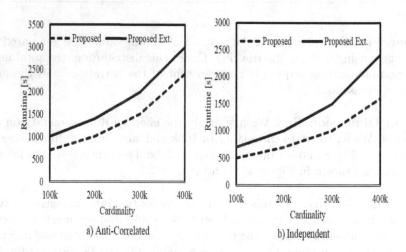

a) Anti-Correlated

b) Independent

Fig. 3. Performance for different cardinality

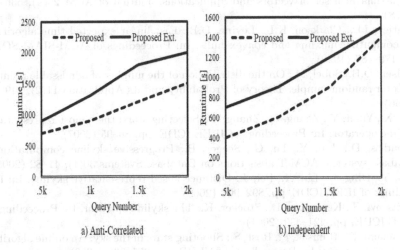

Fig. 4. Performance for Different Query Number

6   Conclusion

This paper addresses k-objects selection problem and give guideline how to selects k objects that are preferable for all users who may have different scoring function. We applied the idea of skyline queries to select the k objects and proposed an efficient algorithm. Experimental results show the effectiveness of the proposed algorithm.

It is worthy of being mentioned that this work can be expanded in a number of directions. First, from the perspective of parallel computing, how to compute k-objects from streaming dataset. Secondly, to design an efficient index based (R-tree/B-tree) algorithm are promising research topics.

Acknowledgment. This work is supported by KAKENHI $(23500180, 25.03040)$ Japan.

References

1. Fagin, R., Lotem, A., Naor, M.: Optimal aggregation algorithms for middleware. In: Proceedings of ACM PODS, pp. 102–113 (2001)
2. Bruno, N., Gravano, L., Marian, A.: Evaluating top-k queries over web-accessible databases. ACM Transactions on Database Systems 29(2), 319–362 (2004)
3. Chang, K.C., Hwang, S.-W.: Minimal probing: supporting expensive predicates for top-k queries. In: Proceedings of ACM SIGMOD, pp. 346–357 (2002)
4. Hwang, S.-W., Chang, K.C.: Optimizing access cost for top-k queries over web sources. In: Proceedings of IEEE ICDE, pp. 188–189 (2005)

5. Bentley, J.L., Kung, H.T., Schkolnick, M., Thompson, C.D.: On the average number of maxima in a set of vectors and applications. Journal of ACM 25(4), 536–543 (1978)

6. Bentley, J.L., Clarkson, K.L., Levine, D.B.: Fast linear expected-time algorithms for computing maxima and convex hulls. In: Proceedings of ACM-SIAM SODA, pp. 179–187 (1990)

7. Nielsen, O.B., Sobel, M.: On the distribution of the number of admissable points in a vector random sample. Theory of Probability and its Application 11(2), 249–269 (1966)

8. Lin, X., Yuan, Y., Zhang, Q., Zhang, Y.: Selecting stars: the k most representative skyline operator. In: Proceedings of IEEE ICDE, pp. 86–95 (2007)

9. Papadias, D., Tao, Y., Fu, G., Seeger, B.: Progressive skyline computation in database systems. ACM Transactions on Database Systems 30(1), 41–82 (2005)

10. Tao, Y., Ding, L., Lin, X., Pei, J.: Distance-based representative skyline. In: Proceedings of IEEE ICDE, pp. 892–903 (2009)

11. Borzsonyi, S., Kossmann, D., Stocker, K.: The skyline operator. In: Proceedings of IEEE ICDE, pp. 421–430 (2001)

12. Kossmann, D., Ramsak, F., Rost, S.: Shooting stars in the sky: An online algorithm for skyline queries. In: Proceedings of VLDB, pp. 275–286 (2002)

13. Chomicki, J., Godfrey, P., Gryz, J., Liang, D.: Skyline with Presorting. In: Proceedings of IEEE ICDE, pp. 717–719 (2003)

14. Tan, K.-L., Eng, P.-K., Ooi, B.C.: Efficient Progressive Skyline Computation. In: Proceedings of VLDB, pp. 301–310 (2001)

15. Vlachou, A., Doulkeridis, C., Kotidis, Y., Vazirgiannis, M.: SKYPEER: Efficient Subspace Skyline Computation over Distributed Data. In: Proceedings of IEEE ICDE, pp. 416–425 (2007)

16. Fotiadou, K., Pitoura, E.: BITPEER: Continuous Subspace Skyline Computation with Distributed Bitmap Indexes. In: Proceedings of DaMaP, pp. 35–42 (2008)

17. Chan, C.-Y., Jagadish, H.V., Tan, K.-L., Tung, A.K.H., Zhang, Z.: On High Dimensional Skylines. In: Ioannidis, Y., et al. (eds.) EDBT 2006. LNCS, vol. 3896, pp. 478–495. Springer, Heidelberg (2006)

18. Tao, Y., Xiao, X., Pei, J.: Subsky: Efficient Computation of Skylines in Subspaces. In: Proceedings of IEEE ICDE, pp. 65–65 (2006)

19. Tao, Y., Lin, W.: XIAO, X.: Minimal MapReduce Algorithm. In: Proceedings of ACM SIGMOD, pp. 529–540 (2013)

20. Park, Y., Min, J, Shim, K.: Parallel Computation of Skyline and Reverse Skyline Queries Using MapReduce. In: Proceedings of VLDB, pp. 2002–2013 (2013)

21. Jiang, D., Tung, A.K.H., Chen, G.: MAP-JOIN-REDUCE: Toward Scalable and Efficient Data Analysis on Large Clusters. IEEE Transactions on Knowledge and Data Engineering 23(9), 1299–1311 (2011)

22. Blanas, S., Patel, J.M., Ercegovac, V., Rao, J., Shekita, E.J., Tian, Y.: A comparison of join algorithms for log processing in MaPreduce. In: Proceedings of ACM SIGMOD, pp. 975–986 (2010)

23. Vernica, R., Carey, M.J., Li, C.: Efficient parallel set-similarity joins using MapReduce. In: Proceedings of ACM SIGMOD, pp. 495–506 (2010)

Implementing the Palomar Transient Factory Real-Time Detection Pipeline in GLADE: Results and Observations

Florin Rusu[1], Peter Nugent[2], and Kesheng Wu[2]

[1] University of California, Merced, CA 95343, USA
[2] Lawrence Berkeley National Laboratory, Berkeley, CA 94720, USA
frusu@ucmerced.edu, {penugent,kwu}@lbl.gov

Abstract. Palomar Transient Factory is a comprehensive detection system for the identification and classification of transient astrophysical objects. The central piece in the identification pipeline is represented by an automated classifier that distinguishes between real and bogus objects with high accuracy. Given that the classifier has to identify the most significant transients out of a large number of candidates in near real-time, the response time it provides is of critical importance. In this paper, we present an experimental study that evaluates a novel implementation of the classifier in GLADE—a parallel data processing system that combines the efficiency of a database with the extensibility of Map-Reduce. We show how each stage in the classifier – candidate identification, pruning, and contextual realbogus – maps optimally into GLADE tasks by taking advantage of the unique features of the system—range-based data partitioning, columnar storage, multi-query execution, and in-database support for complex aggregate computation. The result is an efficient classifier implementation capable to process a new set of acquired images in a matter of minutes even on a low-end server. For comparison, an optimized PostgreSQL implementation of the classifier takes hours on the same machine.

Keywords: parallel databases, multi-query processing, real-time classification.

1 Introduction

The Palomar Transient Factory (PTF) project [1,2] aims to identify and automatically classify transient astrophysical objects such as variable stars and supernovae in real-time. As a secondary objective, a catalog containing the identified transients and other celestial objects is constructed for subsequent querying and analysis. PTF is a comprehensive transient detection system including a wide-field survey camera, an automated real-time data reduction pipeline, a dedicated photometric follow-up telescope, and a full archive of all detected sources (Figure 1 [2]). The computational system supporting the project consists of two separate processing pipelines [2] fed with the images taken by the camera. Between 2000 and 4000 high-resolution (2048 × 4096 pixels) images are taken each night and fed into the two pipelines through high-speed communication links. The total amount of raw data varies between 60 and 100 GB per night. The near-real-time transient detection pipeline [3] has the goal of identifying and classifying transient objects within 30-45 minutes of images being taken. Observation of potential

A. Madaan, S. Kikuchi, and S. Bhalla (Eds.): DNIS 2014, LNCS 8381, pp. 53–66, 2014.

transients by a network of follow-up telescopes is triggered immediately after detection in order to confirm their existence. The main objective of the time-consuming archival pipeline [4] is to create a comprehensive catalog of high-quality images and celestial objects that can be queried using a variety of criteria. It is executed on the entire set of images acquired during one night in order to achieve high accuracy. The execution of the archival pipeline typically takes 4-5 hours [4], but it can extend to several days in some cases [2].

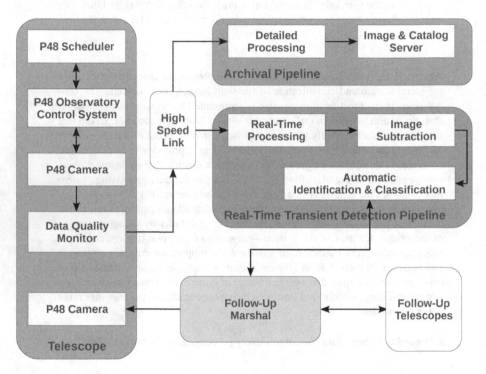

Fig. 1. PTF data flow [2]

Problem Formulation. The problem we address in this paper is the identification of real transient candidates in the near-real-time detection pipeline. Specifically, we focus on the classification phase of the real-or-bogus classifier. The goal of this classifier is to identify real transients with high accuracy. The input consists of a set of candidates extracted during image subtraction and a trained random forest classifier. In the output, the candidates are given scores, i.e., the realbogus score, quantifying the probability of them being real. Only the candidates with realbogus scores higher than a threshold and satisfying a set of additional constraints are considered real.

Contributions. Our objective is to design and implement a real-time classifier capable to keep-up with the continuously increasing size of the PTF repository. Our motivation is the incapacity of the existing PostgreSQL solution to identify transient candidates accurately due to the larger data volumes it has to handle. We present a novel implementation for the real-or-bogus classification in GLADE [5]—a parallel multi-query processing system targeted specifically at analytical workloads. We show how each stage in the classification process is natively supported in GLADE – this is not true for the existing PostgreSQL [6] solution – and prove with experimental results the effect on query execution performance. Since the GLADE implementation reduces the time to investigate a set of candidates to minutes – from hours in PostgreSQL – this allows for considerably more candidates to be thoroughly evaluated, thus increasing the likelihood to find many transients that are otherwise missed by the current solution.

Outline. Section 2 presents in detail automatic transient identification and real-or-bogus classification. It also discusses the PTF solution deployed in production and the problems it has. GLADE is introduced in Section 3. The implementation of the real-or-bogus classifier in GLADE and the results of the experimental evaluation are given in Section 4 which also contains a comparison with two PostgreSQL solutions. We conclude the paper in Section 5.

2   Automatic Transient Identification and Classification

In this section, we present the details of the *automatic detection and classification stage* in the real-time transient detection pipeline (Figure 1 [2]). The input to this stage is represented by the transient candidates extracted during the image subtraction stage. There are in the order of 10^5 such candidates extracted every 45 minutes. Two questions have to be answered for every candidate:

1. Is the candidate real?
2. If real, what is the transient type of the candidate?

Both these questions are answered using automated machine learning classification techniques, i.e., random forest classifiers [3] in this case, that require human intervention only in the follow-up stage. This is necessary considering the number of candidates – 1 to 1.5 million – extracted every night. Since the focus of this work is identifying real candidates – the first question – we present the details and the existing solution in the following. A description of the type classifier can be found elsewhere [3].

2.1   Real-or-Bogus Classification

Any machine learning method consists of multiple phases. First, a series of features have to be defined for the input data. These are used as parameters for the classifier. The features used by the real-or-bogus classifier are extracted during image subtraction and stored in the candidate database. There are 28 features used by the classifier. Second, a training dataset containing labeled examples is used to compute the parameters of the classifier—the training phase. The training dataset consists of 574 candidates manually labeled with the realbogus score by multiple human scanners. At last, the trained classifier is presented with unlabeled examples and the class has to be determined—the classification phase. The output consists of 5 classes, i.e., {bogus, suspect, unclear, maybe, realish}. The probability of a candidate being in each of these classes is returned by the random forest classifier—probabilistic model. The score, i.e., realbogus score, for a candidate is computed as a weighted average of the class probabilities. The final realbogus score takes into account additional information, i.e., the scores of neighboring candidates from the same subtraction. Moreover, a high-scoring candidate is deemed real if and only if it appears in at least two subtractions within 6 days. All the candidates identified as real at the end of the classification phase – 30 to 150 out of 10^5 – are flagged for immediate follow-up and sent to the type identification classifier. We present the exact details of the online classification phase in the following since training is a one-time offline process.

Algorithm 1. *Real-or-Bogus Classification*

Input: new subtraction set (`subtraction`) with corresponding candidate set (`candidate`) and their probabilities (`rb_classifier`) computed by the random forest classifier
Output: a subset of real candidates (`real`)

1. `real` ← IdentifyCandidates(`subtraction`, `candidate`, `rb_classifier`)
2. **for all** r ∈ `real` **do**
3.   **if** SingleAppearance(r, `subtraction`, `candidate`, `rb_classifier`) **then**
4.   `real` ← `real` - r
5.   **end if**
6. **end for**
7. **for all** r ∈ `real` **do**
8.   `ctxt_score` ← CtxtRealBogus(r, `subtraction`, `candidate`, `rb_classifier`)

9.   **if** `ctxt_score` < threshold **then**
10.   `real` ← `real` - r
11.   **end if**
12. **end for**

Identify Candidates. The initial realbogus score – the score returned by the random forest classifier – corresponding to a candidate is computed during image subtraction. It is stored together with other candidate data in the subtraction and candidate database. Candidate identification requires a simple query that returns all the candidates with high realbogus score extracted from subtractions computed during a specified time interval.

Prune Single Appearance Candidates. In order to increase the probability that a high-scoring candidate is indeed real, a candidate identified in the first query has to satisfy an additional condition. The candidate has to appear at a close spatial position in other subtractions close in time to its originating subtraction. Independent of its original realbogus score, a single appearance candidate is pruned away. Since the area and time interval are dependent on the candidate, pruning requires a separate query with different space and time bounds for every candidate. The larger the number of candidates identified in the first query, the more queries have to be executed for pruning.

Compute Contextual Realbogus Score. The final realbogus score of a candidate takes into consideration the score returned by the random forest classifier for the closest k other candidates extracted from the same subtraction. This is called the contextual realbogus score. It is based on this score that the final classification decision is made. Computing the contextual realbogus score is a rather complicated process that involves a nearest-neighbor query followed by a complex aggregate computation. Unlike pruning, the nearest-neighbors are computed only along the spatial dimension, i.e., in the same subtraction. Nonetheless, a separate query has to be executed for every candidate that survives pruning—a considerable problem when the number of candidates is large. Algorithm 1 summarizes formally the stages of the real-or-bogus classification.

2.2   Existing Solution

The current solution implemented in the PTF pipeline is a standard Python [7] application with a PostgreSQL [6] database backend. The database contains 3 tables – `subtraction`, `candidate`, and `rb_classifier` – storing subtractions, candidates, and the scores returned by the random forest classifier. `subtraction` and `candidate` are wide tables having 51 and 46 columns, respectively. Many of the columns are never used in queries. The number of rows in these tables increases continuously as more observations are taken daily. `candidate` and `rb_classifier` already contain a few billion tuples each. Data corresponding to a new set of images are added to the tables during the image subtraction stage. Transient identification and classification execute as database queries. Possible candidates in a given time window are identified with a complex query over the 3 tables. For each such candidate, a time and space nearest-neighbor query is executed from the application to find additional appearances of the candidate. These queries are executed iteratively. For the remaining candidates, another spatial nearest-neighbor query is executed to find close candidates in the same subtraction. The contextual realbogus score is computed in the application by combining the score of the candidate with those of its neighbors. This is another iterative process that goes over the non-pruned candidates one at a time.

Problems with the Existing Solution. The existing approach suffers from a series of problems that make real-time candidate identification and classification in the limited time interval between two subtraction sets – approximately 45 minutes [3] – difficult. Experimental results over a relatively small snapshot of the database from the early stages of the project confirm this problem (Section 5). As the size of the repository increases with the acquisition of new images, the situation will become only worse. As a result, many of the transients are missed simply because there is not enough time to carefully investigate them.

The fundamental limitation is raised by the need to evaluate each candidate sequentially even though the same query template is used across all the candidates. Essentially, two passes over the candidates extracted based on the realbogus score are required to take a decision. And in each pass, a complicated nearest-neighbor query is executed for each candidate. Since the time taken to process one query over the increasingly larger `candidate` table grows continuously with the size of the table, the number of candidates that can be inspected between two subtractions decreases.

The PostgreSQL row-based storage format affects query performance negatively considering the width of `subtraction` and `candidate` tables and the number of attributes used in the query—a small fraction out of the overall number of attributes. While query execution speed can be improved with appropriate indexes, this results in data ingestion time increase due to index maintenance, thus limiting the time available for querying. Essentially, indexing moves the bottleneck from querying to data ingestion.

With the increase in repository size, index maintenance under data ingestion only becomes worse.

Data transfer between the database and the application is another limitation that is a direct consequence of the large number of queries that have to be executed. The reason for this is the lack of support for complex computations inside the database. While user-defined functions (UDF) and user-defined aggregates (UDA) provide extensibility to in-database complex computations, the exclusive SQL invocation limits their applicability. As a result, these complex computations are executed in the application in the current PTF solution.

3   GLADE

Given the aforementioned problems of the existing solution and the incapacity to accurately identify some highly-probable candidates, novel solutions have to be explored. The approach we take in this paper is a novel implementation of the real-or-bogus classifier in GLADE—a parallel data processing prototype we have developed from scratch over the past few years. Although we showed the considerable performance gains GLADE provides over PostgreSQL and Hadoop on a limited set of tasks [5], we have not evaluated GLADE on a complex real-life application yet. Moreover, the characteristics of the PTF real-or-bogus classifier map perfectly on the GLADE architectural features. While these are compelling reasons to carry out such an investigation, the experimental results in Section 5 prove that GLADE is indeed a suitable solution that outperforms the existing PostgreSQL implementation.

GLADE. GLADE [5,8] is a parallel data processing system specifically designed for the execution of analytical tasks specified in SQL enhanced with Generalized Linear Aggregates (GLA). This allows for the execution of a much larger class of analytical computations beyond the standard SQL aggregates. Essentially, GLADE provides an infrastructure abstraction for parallel processing that decouples the algorithm from the runtime execution. The algorithm has to be specified in terms of a clean interface – SQL + GLA – while the runtime takes care of all the execution details including data management, memory management, and scheduling. Contrary to existent parallel data processing systems designed for a target architecture, typically shared-nothing, GLADE is architecture-independent. It runs optimally both on shared-disk servers as well as on shared-nothing clusters. The reason for this is the exclusive use of thread-level parallelism inside a processing node while process-level parallelism is used only across nodes. There is no difference between these two in the GLADE infrastructure.

Architecture. GLADE consists of two types of entities—a coordinator and one or more executor processes (Figure 2). The coordinator is the interface between the user and the system. Since it does not manage any data except the catalog metadata, the coordinator does not execute any data processing task. These are the responsibility of the executors, typically one for each physical processing node. It is important to notice that the executors act as completely independent entities, in charge of their data and of the physical resources. Each executor runs an instance of the DataPath [9] relational execution engine enhanced with a GLA metaoperator for the execution of arbitrary user code specified using the GLA interface.

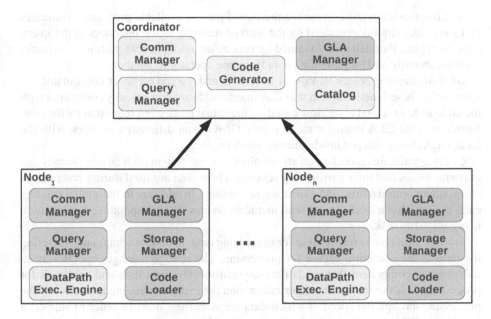

Fig. 2. GLADE architecture

Communication Manager is in charge of transmitting data across process boundaries, between the coordinator and the executors, and between individual executors. Different inter-process communication strategies are used in a centralized environment with the coordinator and the executor residing on the same physical node and for a distributed shared-nothing system. The communication manager at the coordinator is also responsible for maintaining the list of active executors. This is realized through a heartbeat mechanism in which the executors send alive messages at fixed time intervals.

Query Manager is responsible for admission, setup, and query processing coordination across executors and queries. This is a particularly important task since processing is asynchronous both with respect to executors as well as to queries.

Code Generator fills pre-defined M4 templates with macros specific to the actual processing requested by the user generating highly-efficient C++ code similar to direct hard-coding of the processing for the current data. The resulting C++ code is subsequently compiled together with the system code into a dynamic library. This mechanism allows for the execution of arbitrary user code inside the execution engine through direct invocation of the GLA interface methods.

Code Loader links the dynamic library to the core of the system allowing the execution engine and the GLA manager to directly invoke user-defined methods. While having the code generator at the coordinator is suitable for homogeneous systems, in the case of heterogeneous systems both the code generator and the code loader can reside at the executors.

DataPath Execution Engine implements a series of relational operators – SELECT, PROJECT, JOIN, AGGREGATE – and a special GLA metaoperator for the execution of arbitrary user code specified using the GLA interface. They are all configured at runtime with the actual code to execute based on the requested processing. The execution engine has two main tasks—manage the thread pool of available processing resources and route data chunks generated by the storage manager to the operators in the query execution plan. Parallelism is obtained by processing multiple data partitions – chunks – simultaneously and by pipelining data from one operator to another.

GLA Manager executes `Merge` at executors and `Terminate` at coordinator, respectively. These functions from the GLA interface [5] are dynamically configured with the code to be executed at runtime based on the actual processing requested by the user. Notice that the GLA manager merges only GLAs from different executors, with the local GLAs being merged inside the execution engine.

Catalog maintains metadata on all the objects in the system such as table names and attribute names and their partitioning scheme. These data are used during code generation, query optimization, and execution scheduling. In addition to the global catalog, each executor has a local catalog with metadata on how its corresponding data partition is organized on disk.

Storage Manager is responsible for organizing data on disk, reading, and delivering the data to the execution engine for processing. The storage manager operates as an independent component that reads data asynchronously from disk and pushes it for processing. It is the storage manager rather than the execution engine in control of the processing through the speed at which data are read from disk. In order to support a highly-parallel execution engine consisting of multiple execution threads, the storage manager itself uses parallelism for simultaneously reading multiple data partitions.

Range-Based Data Partitioning. Parallel execution is supported in GLADE through data partitioning, i.e., multiple partitions are processed simultaneously by different executors. The tuples of a relation are divided horizontally into chunks containing thousands to a few million tuples. Chunks are stored continuously on disk. The larger the size of the chunk, the longer the size of sequential scans, thus the smaller the number of disk seeks. While tuples can be assigned to chunks in arbitrary order, it is particularly useful for many workloads to have tuples with close values along some attributes grouped together in the same chunk. This corresponds to range-based data partitioning. Generating range-partitioned chunks is typically more complicated since data have to be partially ordered along the partitioning attributes. The benefit is faster execution for range queries since a reduced number of chunks have to be processed.

Column-Oriented Storage. Inside a chunk the columns of a relation are further partitioned vertically, with a disk page storing only values from the same column. Attribute values corresponding to the same tuple are stored at the same relative position inside each column. This allows for immediate tuple reconstruction in memory. The benefit of column-oriented storage is evident in the case of wide relations containing a large number of attributes with only a few of them accessed by every query. When range-based partitioning is combined with columnar storage – the case in GLADE – the amount of data read from disk is minimized since only the chunks and the columns required by the query are retrieved.

Multi-query Processing. GLADE supports concurrent execution of multiple queries by sharing data access across the entire hierarchy—from disk to CPU registers through memory and cache. All the queries reading data from the same relation are connected to a single circular shared scan operator that reads a chunk only once and distributes it to all the queries that require it. While this is standard practice in any multi-query processing system, chunk sharing in GLADE is taken considerably further. Essentially, chunks are shared across all the common operators in the query execution trees corresponding to two queries. This requires merging separate operators with similar functionality into a single mega-operator that combines the operations corresponding to each query. For example, instead of having two selection operators with different predicates – one for each query – a single operator containing both predicates is created in GLADE. The new combined operator is responsible for identifying what queries a chunk is valid for and for setting the correct tuple validity based on the query predicates. The same logic can be applied to other relational operators, including JOIN, GROUP BY, AGGREGATE, and the GLA metaoperator. The code executed by each operator is dynamically generated at runtime based on the running queries.

Complex Aggregates. The GLA metaoperator supports the execution of arbitrary user code specified using the abstract GLA interface [10]. This allows for the execution of complex computations far beyond standard SQL aggregates inside the execution engine without the need to extract data into an application with more powerful computational capabilities. This paradigm shift – bring the code near the data instead of moving data to the code – results in considerable gains especially in the cases where a large amount of data have to be moved.

4 Experimental Evaluation

In this section, we present the GLADE implementation for real-or-bogus classification. We show how data are mapped onto the GLADE storage model, how each task in the classification is expressed as GLADE computations, and how native GLADE features – range-based data partitioning, columnar storage, multi-query processing, and complex aggregate computation – are used in this workload. We provide measurement results that prove a significant improvement over the existing solution and we analyze the reasons for this.

Data. The data we use in the experiments are a snapshot of the subtraction and candidate database. The 3 tables referenced in real-or-bogus classification and their characteristics are given in Table 1. The overall size of the 3 tables when loaded in GLADE is 161 GB. Notice that approximately 5,000 candidates are not classified by the random forest classifier—rb_classifier contains less tuples than candidate. There are 647 candidates per subtraction on average.

The maximum chunk size is fixed at $2^{20} \approx 1$ million tuples across all the tables. This generates a single subtraction chunk and 642 chunks for the other two tables. The size of a full chunk is different though across tables since they contain a different number of columns. Notice that only the columns required in query processing are read for a chunk – not the entire chunk – due to the columnar storage. Thus, even the subtraction table is never read in full unless all the columns are requested by the query. candidate and rb_classifier are range-based partitioned along the subtraction_id attribute. This guarantees that all the candidates extracted from the same subtraction end up in the same chunk. Moreover, candidates from subtractions close in time are also co-located in the same chunk with high probability. This partitioning has two benefits. It minimizes the number of chunks read from disk. And it isolates processing to the chunk level, thus increasing the amount of parallelism achieved by processing multiple chunks simultaneously.

Table 1. Tables used in real-or-bogus classification

Table name	Columns	Rows	Chunks
subtraction	51	1,039,758	1
candidate	46	672,912,156	642
rb_classifier	9	672,906,737	642

Setup. The machine used in the experiments is a low-end server with an Intel Core2 Quad CPU running at 2.66 GHz, 4 GB of memory, and a single 1 TB disk with sequential I/O throughput of 100 MB/s. Ubuntu SMP 10.10 64-bit is the operating system. There is a single GLADE executor in this configuration. It is co-located with the coordinator. Only thread-level parallelism is employed. The DataPath execution engine is configured to use 4 worker threads – one for each core – while the storage manager corresponding to every table assembles 4 chunks simultaneously. The reader might be surprised by our modest system choice given that the PTF pipeline is running in production on a powerful NERSC supercomputer. Nonetheless, our results confirm that even on such a low-end machine GLADE manages to load and classify the candidates in a set of subtractions in less than 20 minutes.

Data Ingestion. The time it takes to ingest the entire dataset depicted in Table 1 in GLADE is 8,222 seconds ($\approx$ 2 hours 15 minutes) out of which 7,056 are spent loading the candidate table. This dataset corresponds though to many nights of observations. To determine how long it takes to ingest a set of subtractions generated at one instance in time, we chose a random night in the dataset, i.e., the night of October 10, 2011, compute its corresponding statistics, and then extrapolate the loading time based on these statistics. There are 2,997 subtractions taken during this night and 1,939,059 candidates at an average rate of 647 candidates per subtraction. The time taken to ingest these data into GLADE is only 24 seconds. Considering that the ingestion is distributed over 10 periods of 45 minutes each, the average ingestion time for a set of subtractions is less than 3 seconds.

Candidate Identification. The first stage in real-or-bogus classification is to identify candidates with high realbogus score assigned by the random forest classifier. The corresponding query (1) contains a 3-way join between subtraction, candidate, and rb_classifier and a series of selection predicates on each of the tables. The most important predicate is a range selection on subtraction that limits the search to the images acquired most recently. Due to range-based partitioning and columnar storage, in the GLADE implementation this query reads only the chunks and columns that generate results. In the optimal situation, a single chunk is processed from each of the 3 tables. Out of the almost 2 million candidates detected during the night of October 10, 2011, only 40,087 are classified as real by the random forest classifier. This is only 2%. The number can be easily increased by relaxing the conditions in the query. It takes GLADE only 9.2 seconds to find the real candidates, i.e., 0.92 seconds per subtraction.

```
SELECT s.ujd, c.sub_id, c.id, c.ra, c.dec,
       c.xint_new, c.yint_new, c.pos_sub
FROM
  subtraction s JOIN candidate c ON (c.sub_id = s.id)
  JOIN rb_classifier rbc ON
    (rbc.sub_id = c.sub_id AND rbc.candidate_id = c.id)       (1)
  WHERE s.ujd > 2455844 AND s.ujd < 2455845 AND
  rbc.realbogus > 0.17 AND rbc.bogus < 0.35 AND
  c.b_image > 0.7 AND c.pos_sub = 'True' AND
  (c.a_image < 3.0 OR c.mag < 15.0)
```

Candidate Pruning. Each of the candidates identified by the random forest classifier is further checked before deemed real. The first condition a candidate has to pass is that it appears in more than one subtraction at a position close to the original position where it was first spotted. This is expressed as a complex nearest-neighbor query (2) along the space and time attributes. In the current implementation of the PTF pipeline, one such query is executed sequentially for every candidate. This is 40,087 queries for our example night or approximately 4,000 queries for every subtraction set. Since all these queries have to be executed in less than 45 minutes, this step is by far the bottleneck of the entire process. GLADE multi-query processing kicks in perfectly in this situation thus allowing for multiple candidates to be checked at the same time. Most importantly though, the time to check many candidates – up to 64 candidates in the current GLADE implementation – is the same as checking a single candidate. The reason for this is that the queries have identical execution plans – more or less some constants – which allows for maximum data access sharing. When coupled with range-based partitioning and columnar storage, it takes only 18 minutes to check the 4,000 candidates identified in a subtraction set—18 seconds for a group of 64 candidates. 560 candidates survive pruning on average for each subtraction set.

```
SELECT COUNT(*)
FROM
  subtraction s JOIN candidate c ON (c.sub_id = s.id)
  JOIN rb_classifier rbc ON (rbc.candidate_id = c.id)
  WHERE c.ra BETWEEN (%1f,%2f) AND                             (2)
  c.dec BETWEEN (%3f,%4f) AND
  (s.ujd BETWEEN (%5f,%6f) OR s.ujd BETWEEN (%7f,%8f))
    AND (rbc.realbogus > 0.07 OR c.pos_sub <> 'True') AND
  c.b_image > 0.7 AND (c.a_image < 3.0 OR c.mag < 15.0)
```

Contextual Realbogus Computation. For the surviving candidates, the contextual realbogus score is computed based on the probability of being real of their nearest-neighbor candidates in the subtraction. This requires another iterative process in which each surviving candidate is examined independently. The difference from pruning is that the contextual realbogus score cannot be computed inside the database. Instead it is computed in a Python script that extracts the necessary data from the database using query (3). This is not required in GLADE though since complex aggregates can be expressed as GLAs and executed inside the system without moving data between processes. In addition to the savings in execution time, the GLA mechanism allows for all the computation to be confined to the database engine—a cleaner and easier to understand solution. In GLADE, the contextual realbogus score for the 560 candidates surviving pruning in a subtraction set is computed in 88 seconds—it takes 10 seconds on average to compute the score for a group of 64 candidates.

Table 2 summarizes the results we obtained for processing the October 10, 2011 data in GLADE. These are average results for processing a subtraction set. The overall time to classify the candidates is less than 20 minutes. This is less than half the length of the interval between two sets of images are ingested, i.e., 45 minutes. The remaining time can be used either to increase the rate at which images are ingested or to analyze more candidates—some of the conditions based on which candidates are pruned are arbitrary and they are targeted at reducing the overall classification time. This is not a problem in the GLADE implementation though.

```
SELECT c.id,
```
$$\sqrt{(\texttt{c.xint\_new-\%1f})^2 + (\texttt{c.yint\_new-\%2f})^2} \texttt{ AS dist}$$
```
FROM
  subtraction s JOIN candidate c ON (c.sub_id = s.id)        (3)
  JOIN rb_classifier rbc ON
  (rbc.sub_id = c.sub_id AND rbc.candidate_id = c.id)
WHERE s.id = %3i AND c.pos_sub = '%4s'
ORDER BY dist
```

PostgreSQL Solutions. In order to compare the proposed GLADE approach with the existent solution, we devise two alternative PostgreSQL databases. The first database does not contain any optimizations. There are no indexes or any other structures for enhancing query performance. The second database defines indexes for all the attributes used in selection predicates or join conditions across the three workload queries. This is the solution implemented in the PTF production pipeline. We deploy these two databases on a PostgreSQL 8.4 server running on the same test machine. We modify the server configuration in order to maximize usage of the available memory resources in the system, e.g., we set shared_buffers to 3 GB.

Table 2 contains the results for the two PostgreSQL database implementations. The indexed implementation outperforms the non-indexed version considerably at query processing. The gap is as much as 6 orders of magnitude for the contextual realbogus computation. The reason for this is that the non-indexed database has to read all data from all the tables in order to perform any query. Since no indexes are available, sequential scan is the only feasible path access strategy. Indexes reduce dramatically query execution time since the tuples satisfying the highly-selective predicates can be identified with as little as a single disk access. Due to the large buffer pool, disk access is not even required at all in many situations. Indexes also play an important role in the selection of the join algorithms used in query execution plans. The situation is radically different though for data ingestion. While it takes less than a minute to load a new set of candidates in the non-indexed database, it takes 68 minutes to do so in the indexed version. This 61 factor difference is entirely due to index maintenance. Adding tuples to the `candidate` table requires insertions in each of the 9 indexes defined over it. Although this might not seem such a difficult problem at first, in the case of a batch of 200,000 insertions the probability to encounter some time-consuming index reorganizations is quite high.

Overall, none of the PostgreSQL solutions meets the requirement to ingest and identify a new set of candidates in less than 45 minutes. The index-based solution deployed in production takes 75 minutes on our test machine—out of which 60 minutes are spent for data ingestion. The non-indexed version is far from this requirement. A possible solution to decrease the loading time for the indexed database is to reduce the number of indexes. The expectation is that the decrease in loading time offsets the increase in query execution time and for some combination of indexes the overall time drops below 45 minutes. Finding the optimal index combination is a hard problem that requires the investigation of an exponential number of alternatives.

Table 2. Average results for processing a subtraction set on October 10, 2011

Phase	GLADE	PostgreSQL	PostgreSQL + indexes
Data ingestion	3 sec	59 sec	1 hour 8 sec
Identification	0.92 sec	45 sec	4.67 sec
Pruning	18 min	607 hours	15 min 39 sec
Contextual realbogus	1 min 30 sec	68 hours	0.79 sec

Observations. When we compare the proposed GLADE solution to the PostgreSQL indexed database, we remark some interesting aspects. Overall, GLADE outperforms PostgreSQL by a factor of 3.88. This is entirely due to the efficient GLADE data loading mechanism which is faster by 3 orders of magnitude. Since GLADE does not employ any secondary data structures to enhance query performance, it is not as efficient as indexed PostgreSQL in answering queries. The difference between the two systems – only 24% – is considerably smaller when compared to the basic PostgreSQL implementation. The GLADE architecture specifically targeted at analytical processing and optimized for read-mostly workloads is responsible for providing similar query performance to indexed PostgreSQL but without the associated increase in database size – the indexed PostgreSQL database is twice as large as GLADE – and ingestion time—GLADE loads new candidates a factor of 60 faster. Range-based partitioning and columnar storage combine together in order to minimize the amount of data read from disk across all types of range queries. Dedicated support for the execution of any user code inside the system eliminates data movement almost entirely and allows for complex computations to be executed right near the data. For the PTF workload though, the most significant gains are due to multi-query processing. Instead of verifying each candidate one at a time, GLADE allows for up to 64 candidates to be evaluated simultaneously in the same amount of time. This is because all the queries have identical execution plans and GLADE is capable to combine them into a single plan in which the operators share access along the entire data path—from disk to CPU registers through main memory and cache.

5   Conclusions

In this paper, we present a novel implementation for the real-or-bogus classification in GLADE—a parallel multi-query processing system targeted specifically at analytical workloads. We show how each stage in the classifier – candidate identification, pruning, and contextual realbogus – maps optimally into GLADE tasks by taking advantage of the unique features of the system—range-based data partitioning, columnar storage, multi-query execution, and in-database support for complex aggregate computation. The result is an efficient classifier implementation capable to process a new set of acquired images in a matter of minutes even on a low-end server. For comparison, the existing optimized PostgreSQL implementation of the classifier is a factor of 3.88 slower. Due to this reduction in the time to investigate a set of new candidates, considerably more candidates can be thoroughly evaluated, thus increasing the likelihood to find many transients that are otherwise missed by the current solution.

References

1. Palomar Transient Factory (November 2013),
   http://www.astro.caltech.edu/ptf/
2. Law, N.M., et al.: The Palomar Transient Factory: System Overview, Performance and First Results. CoRR abs/0906.5350 (2009)
3. Bloom, J.S., et al.: Automating Discovery and Classification of Transients and Variable Stars in the Synoptic Survey Era. CoRR abs/1106.5491 (2011)
4. Grillmair, C.J., et al.: An Overview of the Palomar Transient Factory Pipeline and Archive at the Infrared Processing and Analysis Center. In: Astronomical Data Analysis Software and Systems XIX. ASP Conf. Ser., vol. 434, pp. 28–36 (2010)
5. Cheng, Y., Qin, C., Rusu, F.: GLADE: Big Data Analytics Made Easy. In: Proceedings of 2012 ACM SIGMOD International Conference on Management of Data, pp. 697–700 (2012)
6. PostgreSQL, http://www.postgresql.org/ (November 2013)
7. Python Programming Language (November 2013), http://www.python.org/
8. Cheng, Y., Rusu, F.: Astronomical Data Processing in EXTASCID. In: Proceedings of 2013 SSDBM Conf. on Sci. and Stat. Database Management, pp. 387–390 (2013)
9. Arumugam, S., Dobra, A., Jermaine, C., Pansare, N., Perez, L.: The DataPath System: A Data-Centric Analytic Processing Engine for Large Data Warehouses. In: Proceedings of 2010 ACM SIGMOD International Conference on Management of Data, pp. 519–530 (2010)
10. Rusu, F., Dobra, A.: GLADE: A Scalable Framework for Efficient Analytics. Operating Systems Review 46(1), 12–18 (2012)

Exploratory Analysis of Light Curves: A Case-Study in Astronomy Data Understanding

Aditi Mittal[1], Abhishek Santra[2], Vasudha Bhatnagar[3], and Dhriti Khanna[4]

[1] Hans Raj College, University of Delhi, Delhi, India
mittal.aditi@hotmail.com
[2] Independent Researcher
abhishek.santra@gmail.com
[3] South Asian University, New Delhi, India
vbhatnagar@cs.sau.ac.in
[4] Acharya Narendra Dev College, University of Delhi, Delhi, India
dhriti0610@gmail.com

Abstract. Data acquisition in Biology and Astronomy has seen unprecedented growth in volume since the turn of the century. It will not be an exaggeration to state that the needs of these two sciences are pushing computer science research to new frontiers. The focus of this paper is astronomy, which since inception of Virtual Observatory and commissioning of massive sky surveys is gasping for knowledge in data deluge.

Astrocomputing, which subsumes Astroinformatics, is a recent multidisciplinary field of research with computer science and astronomy at the core. In this article we dwell upon the opportunities and challenges for machine learning and data mining research thrown open by this emerging discipline. We present a case study of an ongoing work on exploratory analysis of unclassified light curves. Though scientific analysis and interpretation of the results of the study are pending, the exercise demonstrates the merit of customized exploratory approach for study. The approach is general and can be applied to light curves obtained from any survey. Owing to the gargantuan scale of astronomy data processing requirements, we discuss scalability of the proposed method.

1 Introduction

With the dawn of *Digital Astronomy*, computing technologies and methods have rapidly gained attention of astronomers and astrophysicists [34]. Vision for intense collaborations for acquisition, sharing and analysis of astronomy data led to development of Virtual Observatory [1]. The movement has resulted into development of astronomy datasets and resources by more than a dozen groups, striving towards the goal of advancing astronomy research using sophisticated computing technologies. The previous decade has also seen commissioning of several different types[1] of sky surveys of varying scale that have resulted into

---

[1] Optical surveys (Sloan Digital Sky Survey [2], Catalina Realtime Transient Survey [3], Palomar Transient Factory [4]); Infrared surveys (Akari [5], WISE [6]); Radio surveys; Multiwave-length surveys etc..

A. Madaan, S. Kikuchi, and S. Bhalla (Eds.): DNIS 2014, LNCS 8381, pp. 67–94, 2014.
© Springer International Publishing Switzerland 2014

accumulation of astronomy data as never before, in the history of mankind. Number of discoveries made from these data is astounding, while those waiting to happen is beyond imagination.

Large Synoptic Sky Survey (LSST) [7] and Square Kilometer Array (SKA) [8] are some of the planned surveys in advanced stage of execution, which have created excitement and awe in equal proportions among astronomy community. LSST alone is expected to spew petabytes of data per night for more than a decade starting in 2020. SKA is also a global science project in the area of radio astronomy, unprecedented in size. In this scenario of data acquisition on humongous scale, astronomy community is aggressively gearing up to store, manage, share and analyze mammoth volumes of data with the help of state-of-the-art information technology.

Computing technologies provide foundational setting for successful achievement of goals and aspirations of this scientific community [35]. Virtual observatory project, which is an ecosystem of "mutually compatible datasets, resources, services, and software tools that use a common set of technologies and a common set of standards" [1], is a fine illustration of dependence of ambitious astronomy survey projects on information technology. Sloan Digital Sky Survey, designed in previous century aimed to collect, process and distribute vast amounts of photometric and spectroscopic data. The scale at which the project was planned, could not have been possible without sophisticated database technology accompanied with the compute power available then, even though all the other necessary technical developments (detectors, optics etc.) were in place [9]. All images, spectra, and measurements recorded so far in SDSS are public and available on-line with an object-oriented operational database system as bedrock.

Upcoming surveys are remarkably demanding in terms of technological infrastructure because of the scale [7,8]. Needs for high-speed data acquisition, their communication to distributed storage, real-time constraints for data retrieval and processing have strained and jolted several branches of computer science and engineering to accelerate and advance in accordance. Similar requirements that have emerged in other sciences facing data deluge (e.g. biology and geology) are pushing computer science and engineering disciplines to new frontiers. In this paper we focus on the challenges and opportunities presented by astronomy and astrophysical sciences.

Astrocomputing is the term that has been used in recent times, to connote encompassing aspects of computer science and engineering that contribute towards accomplishment of objectives and ambitions of large sky surveys. Figure 1 presents the canvas of Astrocomputing showing different areas of computer science and engineering, which coherently and seamlessly charade to provide a solid platform to do science.

Foundational framework to capture raw signals from hardware (telescope and instrument control software) and perform on-site processing under severe time constraints, is provided at the lowest layer of the stack by embedded and real-time systems. Typically, "there is a hierarchy of ever more distilled and value added data products, starting from the raw instrument output and ending with

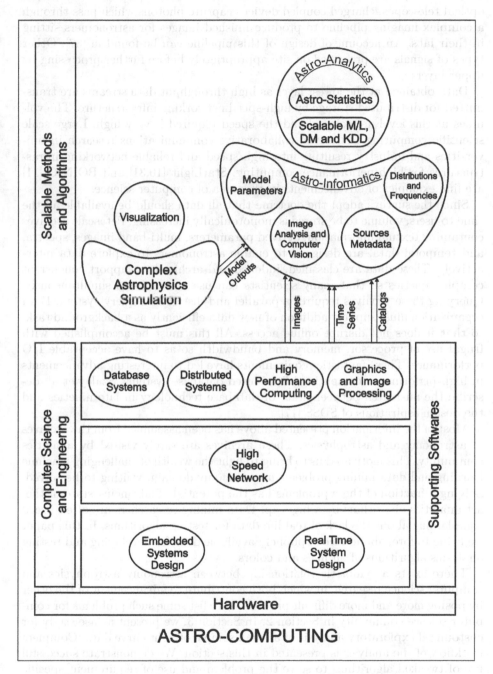

Fig. 1. Canvas of Astrocomputing: For cause of Digital Astronomy

ever more sophisticated descriptions of detected objects" [42]. For example, in optical telescopes, charged coupled devices capture photons which pass through a complex imaging pipeline to produce finished images for astronomers sitting in their labs. An account of design of this pipeline can be found in [37]. Other types of signals are processed on site appropriately before further processing by upper layers.

Data obtained at the lowest layer as high-throughput data streams are transmitted for distributed storage by high-speed networking infrastructure. The volumes at this level are colossal and the speed required is very high. Large scale scientific computing has driven collaborative communications research in universities and industry resulting into high speed and reliable networking infrastructures for high-performance computing. StarLight [10,31] and BOINIC [11] are fine examples of advancements in this area of computer science.

Since the surveys adopt the doctrine that all data should be available all the time to users, volume to be stored is monotonically increasing. Petascale archives containing textual information, derived parameters, multi-band images, spectra, and temporal data, are designed to enable astronomers to explore data interactively. These data are classified, indexed and archived to support concurrent complex queries so that many scientists can use the archives simultaneously. Querying these archives requires a parallel and distributed query system. Data organization must support addition of new data efficiently as a background task, so that it does not disrupt online access. All this must be accomplished with frugal use of processor, memory and bandwidth so as to have acceptable I/O performance. These fantastic requirements have led to interesting advancements in high-performance computing, distributed database systems. Szaley et al. describe the mismatch between state of database technology in late nineties and the design aspirations of SDSS [17].

Most of the information presented above has been assembled from the archives of astronomy and astrophysics. These archives are rarely visited by analytics community. This motivated us to bring to light the wealth of challenging machine learning and data mining problems in astronomy domain, waiting to be solved. Solving a fraction of these problems has the potential of advancing state-of-the-art analytics algorithms by a big leap. Data mining researchers are often heard complaining about the lack of real life data for testing algorithms. In this paper we bring to fore, abundance of publicly available data for developing and testing analytics algorithms of all hues and colors.

There exists a symbiotic relationship between astronomy/astrophysics and computer science research, in which both contribute to advancement of the other by posing more and more difficult problems. We list some such problems for computer science community in Section 2. In Section 3, we present a case study for customized exploratory analysis of a small sample of light curve data. Complete workflow of the analysis is presented in this section. We demonstrate successful use of tweaked algorithms to solve the problem and use of requirement specific quality metrics. In Section 4, we explain the scalability of the workflow, and finally conclude the paper in Section 5.

2   Opportunities and Challenges for Analytics

Intense collaboration between Computer Science and Astronomy researchers offers first-class opportunities to both communities to advance sciences of their respective interests. Particularly, computer science researchers are in unique vantage position to contribute theory and methods not only in their own field, but also to other sciences. Historically, new frontiers of computer science were explored when the current technology fell short of meeting the expectations of other domains.

Astrocomputing provides computer science researchers a fertile ground for exciting problems not only in machine learning, but also prominently in high-performance computing and networking, databases, information retrieval, distributed computing, visualization and image processing. Here we restrict our discussion to astroanalytics, which circumscribes a wide variety of techniques in machine learning, data mining, visualization and statistics (Astrostatistics) etc. Astroanalytics is enabled by Astroinformatics, which includes study and practice of data models, data transformation and normalization methods, indexing techniques, information retrieval and integration methods, content-based and context-based information representations, consensus semantic annotation tags, taxonomies, ontologies, and more [34]. Note that we distinguish between astroanalytics and astroinformatics, even though sometimes the line dividing them blurs. For example, discovering ontology for astronomy may itself require data mining techniques, while some data mining algorithms may use ontologies for discovering knowledge. Distinguishing between the two sub-areas help computer science researchers to define their research goals lucidly. Ball and Brunner present an extensive survey of data mining and machine learning techniques with applications in astronomy [18]. The survey holds as much value for analytics community as for astronomers, because of the clear elucidation of analytics challenges.

Data preparation is the foremost challenge in astroanalytics. Catalogues store variety of data which need to be merged, cleaned and transformed according to the goals set by astronomers solving the science problem. Features of the objects that may aid the analysis and accomplish the goals are often unknown to the scientists. Guided by earlier scientific works, astronomers strive to determine the relevant set of attributes for several important classification tasks. Necessity of principally decisive features for success of classification algorithms does not need advocacy. *Therefore, development of feature selection and construction algorithms that embed domain knowledge is the compelling need of astroanalytics.*

Extending the point to data mining algorithms, we note that current algorithmic designs for analytics are also not sufficiently mature to capture desired domain knowledge. Intuition for scientific analysis and interpretation is not captured by majority of the algorithms belonging to the wide spectrum of analytics techniques. This constitutes a generic challenge, and progress in development of mechanisms to implant domain knowledge in algorithms will serve all spheres of data analytics. *Domain-aware* data preparation algorithms will augment the power of *domain-aware* analytics algorithms. Some early works proposed use of ontologies during data mining [39,40,36], but the community has paid less than desired attention to this area. *This direction needs to be re-examined and the impediments for embedding domain knowledge in data mining algorithms need to be identified and overcome.*

Interactivity and iterative nature of KDD process is crucial for its success. Evaluation of the feature set being tested for classification, cleaning of data, choice of data mining algorithm etc. have substantial impact on the quality of discovered patterns. In the current environment for analytics, it is only towards the end of the data mining step that the analyst/scientist realizes the merit or weakness of the previous steps. By then much expense has been incurred in terms of time, computation and human resource. The problem is aggravated gravely when *big data* are being analyzed [29]. An interactive environment which permits users to peek into intermediate stages of each KDD process step to judge whether the progress is as per expectation, is high on the wish list of astro-analysis enthusiasts. *Developing environments for interactive data analysis is another opportunity presented by burgeoning needs of astroanalysis.* Challenges for development of such environments have been deliberated in [29]. Chen et al. characterize MapReduce workloads from commercial domain, which are driven in part by interactive analysis, and which make heavy use of query-like programming frameworks on top of MapReduce [22]. Similar characterizations for astronomy related problems will be extremely useful for astroanalysis.

Visualization is a powerful technique to uncover patterns hidden in data [33]. It involves mapping data attributes to visual properties such as position, size, shape, and color etc. so as to make sense out of data, which is particularly propitious in case of *big data*. Recent advances in interactive visualization have been surveyed in [43]. Heer et al. present a taxonomy of interactive dynamics for visual analysis [30]. However, heavy simulations running on supercomputing facilities quickly catch-up with the advances made in visualization. Powerful simulations of cosmic evolution (star, planet, galaxy etc.) in astrophysics often depend on effective visualizations and always insist for more from visual analytics community. Interactivity is highly desirable in visual exploration of astronomy datasets because repeated explorations help users to develop insights about significant relationships and causal patterns in data, which accelerates progress. *Large astronomy and complex simulation datasets present a challenge for interactive visualization softwares and human-computer interaction.*

Mining of high-speed data streams is a well researched area in data mining for more than a decade [25]. High-throughput streams arising from the data acquisition layer are important source of hidden patterns that evolve with time. Most important task in these streams is to monitor evolution and report anomalous behaviour. Novelty detection algorithms and monitoring of astronomy data streams have drawn limited attention. We believe that algorithms for mining astronomy data stream will employ extreme domain knowledge as noticeable in [19]. Though algorithms for outlier detection, clustering and classification of data streams form an extremely rich set of data mining tools [15,23], hardly a few have been designed specifically for astronomy domain. Some that must be noted for their domain specific design are [41,26,19]. *Computer science challenge here is to adapt some effective data stream mining algorithms to match the speed of the astronomy data streams, and develop novel algorithms with ingrained domain knowledge.*

Commonly used metrics for discovered patterns do not sometimes conform to the needs of astronomy community. For example, commonly used metrics for cluster quality like mean squared error, cohesion, separation etc., are not sufficient for an astronomer. In presence of high degree of uncertainty in data these metrics are not reliable. Instead attribute wise means, standard deviations, distributions are more informative for a scientist. This is so, because each attribute has a physical interpretation and its value is indicative of some scientific reason for that data value. *Reconciling evaluation metrics for different types of patterns will have far-reaching influence in design of domain-aware algorithms.*

Uncertainty and missing values too pose serious challenge to analytics algorithms. Errors during data acquisition, approximations due to low level processing are inevitable in astronomy data. Development of robust algorithms to handle noise and missing data is imperative to solve astronomy problems.

Some of the challenges mentioned above are also present in commercial and other science domains. Sooner than later, these problems will be solved by analytics researchers by developing appropriate algorithms and methodologies. It is only that the astronomy community is in a hurry to solve the science problems before it drowns in data, and has startled the computer science community into action.

3   Exploratory Data Analysis of Light Curves

In this section we describe an ongoing work on exploratory data analysis to elucidate the customized methodology that was followed to understand light curve data. This work is in progress and the scientific analysis and interpretation of the results are pending. The work is a case study of tweaking and augmenting existing data mining algorithms to solve the given astronomy problem. We justify the decisions and actions at each step of the case study.

3.1   Dataset Description

Dataset is obtained from Catalina RealTime Transients Survey [12]. This survey is designed to capture rare and interesting transient phenomena for astronomical exploration. It covers around thirty three thousand square degrees of the sky. The survey detects and publishes all transients within minutes of observation [3]. The survey has accelerated discovery and classification of transients in recent years and has invoked deep interest among astronomers in time domain astronomy [14].

The data set consists of light curves of about 1720 objects with unknown types. A *light curve* of a celestial object is a sequence of observations of the flux received from the object. Essentially, it is a kind of time-series data with a peculiar stipulation that for each object the number of observations may vary, and so may the time period between observations. Each observations has four attributes viz. masterId (uniquely identifies the object), mag (magnitude of the light flux), magerr(error in magnitude) and ra and dec (right ascension and declination determine position of the object in the sky).

Sample light curves for two objects are shown in Figure 2 and Figure 3. Note that for object 1 there are only five observations, while for the second object the number of observations is more than 400. Further, time period (MJD column) between the observations is irregular. This happens because the observations correspond to timings when the object is visible.

Subfigure 2a and 3a show two sample sets of measured flux observations. Subfigures 2b and 3b show their respective plots with error bars, emphasizing the uncertainty in data. These characteristics of the light curves make them resistant to conventional treatment of time series. Handling of light curves offers *computer science challenge to handle missing data/observations and uncertainty.*

Using domain knowledge available in [13], data transformation was performed and each object was represented by the following ten statistical measures derived from the light curves.

    i. *beyond1std:* Percentage of points beyond one standard deviation from the weighted mean

    ii. *flux_percentile_ratio_mid20:* Ratio of flux percentiles (60th - 40th) over (95th - 5th)

    iii. *flux_percentile_ratio_mid50:* Ratio of flux percentiles (75th-25th) over (95th-5th)

    iv. *linear_trend:* Slope of a linear fit to the light-curve fluxes

    v. *median_buffer_range_percentage:* Percentage of fluxes within 10% of the amplitude from the median

    vi. *pair_slope_trend:* Percentage of the last 30 pairs of consecutive flux measurements that have a positive slope

    vii. *percent_amplitude:* Largest percentage difference between either the maximum or minimum flux and the median

    viii. *skew:* The skew of the magnitudes

    ix. *small kurtosis:* Kurtosis of the magnitudes, reliable down to a small number of epochs

    x. *magratio:* Indicates whether the object spends most of its time above or below the median

Data thus prepared was further subjected to cleaning, and possible artifacts were removed using domain knowledge. After this step, 1709 objects remained in the dataset. *Using deep domain knowledge to prepare the data set is a serious computer science challenge.* Same data may be transformed and cleaned differently depending on the analytics goals that are set by the astronomy collaborators. Designing and developing suitable and scalable techniques for this purpose lies completely in purview of computer science.

```
#MasterID,Mag,Magerr,MJD
1009071029268,20.42,0.36,53742.52406
1009071029268,20.39,0.36,53742.52895
1009071029268,20.48,0.36,54115.49006
1009071029268,19.72,0.25,54591.29065
1009071029268,20.64,0.40,56073.15876
```

(a) Periodic Observations of Object 1

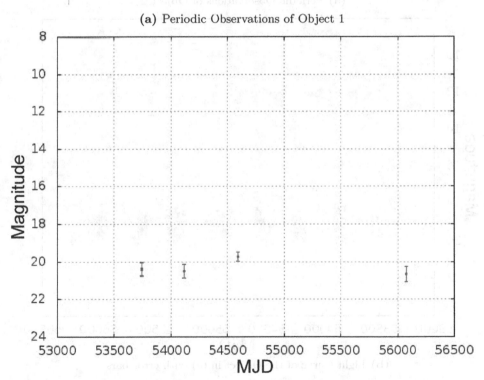

(b) Light Curve of the Object in (a) with error bars

Fig. 2. Light curve illustrations 1

```
#MasterID,Mag,Magerr,MJD
1121068005226,18.48,0.14,53469.35511
1121068005226,19.04,0.17,53469.36320
1121068005226,19.60,0.23,53469.36320
.
.
.
1121068005226,19.49,0.21,56089.25203
1121068005226,18.74,0.15,56089.25818
1121068005226,19.88,0.26,56089.25818
```

(a) Periodic Observations of Object 2

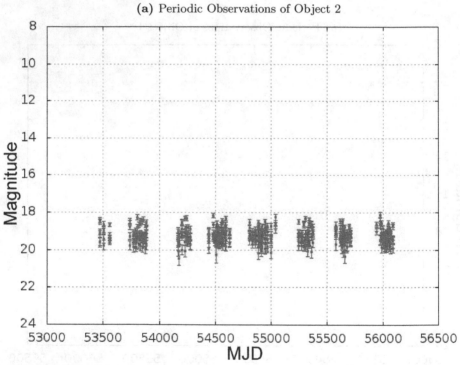

(b) Light Curve of the Object in (a) with error bars

Fig. 3. Light curve illustrations 2

3.2 Methodology Adopted

Recall that the goal of analysis is to explore light curve data for identification and possible classification of objects separating the transients from the non transients. The data were unlabelled and we did not have any idea of identifying the feature distinguishing the two. In this situation we decided to group the data based on inherent similarities and proceeded with clustering as the first step.

We decided to use grid based clustering because of high data uncertainty that prevails. Use of distance function in presence of errors is likely to accentuate data discord and deteriorate the quality of clustering. Grid based clustering methods discretize data space into regions called cells at user defined granularity (g), and locate data instances in these cells. Subsequently, topologically adjacent cells with sufficient number of points are coalesced to reveal clusters in data [28]. Connected Component Analysis is used to coalesce populated adjacent cells to discover clusters. This algorithm is generic and simplified form of effective grid based clustering algorithm [16]. Interested reader is referred to [21] for the algorithm and computational details of adjacency.

The clusters thus found are manifestation of unknown physical phenomena. Each cluster is indicative of the member objects sharing similar physical properties, observed properties and the underlying scientific circumstances. Thus clustering creates a discriminative categorization of data. Since the clustering results heavily depend on the granularity of the grid, determining the *right* granularity is often a non-trivial problem for a novice user. In the context of astronomy, we envisage that for data acquired from a particular source it is possible to ascertain suitable value of g once and use it for subsequent analysis.

Each cluster thus obtained is assigned a unique label to identify its category. Often, heterogeneity within the class is not apparent. Hence it is interesting to examine if sub-categories exist in each category (class). Deeper analysis of labeled data may unveil micro-distributions (μds), discovery of which may lead to an improved understanding of the domain.

To discover micro-distributions from each cluster we used an advanced grid based algorithm [20]. DUSSC (Data Understanding using Semi-Supervised Clustering) is an incremental algorithm designed to discover heterogeneity in a class by discovering hidden micro-distributions. The algorithm uses a measure called *Maximal Information Coefficient(MIC)*, which quantifies the degree of influence of each attribute on the class label. MIC values are used as weights while computing distances between data points. It accepts a parameter r from user that signifies the resolution at which the data is desired to be analyzed. The algorithm locates data instances in grid cells while taking into consideration the class labels, and clusters using connected component analysis. Data instances with alien class labels in a cell are branded as outliers. Sparsely populated cells or data instances far off from core regions are also segregated. All this analysis is performed using thresholds derived from r. Micro-distributions in each class are reported for each cluster (class) separately. The details of the algorithm are skipped in the interest of space and are available in [20].

At this stage, some explanation regarding use of distance function is warranted. We avoided distance based clustering algorithm in favor of grid clustering for the starting dataset because of the inherent uncertainty in observations. Most of the noisy data gets filtered out as we chose cluster with significant membership for later analysis. While investigating clusters, we are more confident of the improved data quality and significantly reduced uncertainty. Hence at this stage we do not hesitate to use weighted distance function. Recall that the weights capture the influence of attributes on the class labels. Our hunch is vindicated by the results as will be clear in the next section.

Figure 4 summarizes the adopted work-flow for this analysis.

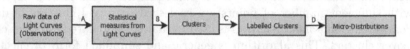

Fig. 4. Work-flow of the analysis of light curve data; A. Cleaning and Data Transformation based on domain knowledge, B. Grid based Clustering (parallel / sequential), C. Labeling of Clusters, D. Investigating clusters for heterogeneity

3.3   Results and Discussion

The data set consisting of 1709 objects was processed at granularity $g = 20$ by the clustering algorithm. Three significant clusters were obtained containing 24, 15 and 1008 objects. We decided to ignore remaining data instances as they were distributed over large number of clusters of sizes 1 - 4. Clusters were labelled to signify respective classes. Each cluster is expected to be representative of an unknown physical phenomenon responsible for the generation of the data. To explore the level of heterogeneity within the clusters (class) i.e. find out how cohesive the clusters were, DUSSC algorithm was applied. At optimal resolution of $r = 0.45$, we obtained the following results

i. **Cluster 1 (24 Objects)**: This cluster of 24 objects threw-up three micro-distributions of 3, 5 and 11 objects. The remaining 5 objects were regarded as outliers as either singular in the cell or far off from the three micro-distributions. Light curves of the objects in each micro-distribution are plotted in Figure 5 to Figure 7. In the first plot the three objects have different magnitudes but have strikingly similar variability trend with small error bars. The second plot shows more variability than the first plot and the objects have bigger error bars. The third plot shows that these objects have highest variability and still bigger error bars.

ii. **Cluster 2 (15 Objects)**: This cluster of 15 objects revealed two micro-distributions of 6 and 9 objects. No outliers were found from this cluster. The light curves of the objects corresponding to the two micro-distributions are plotted in Figure 8 and Figure 9. Objects in the first micro-distribution

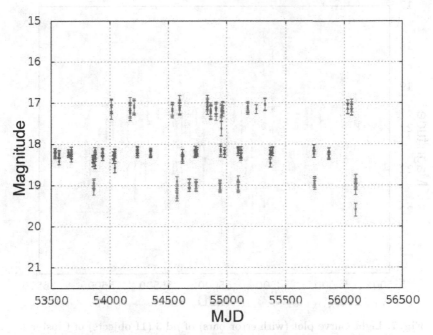

Fig. 5. Light Curve plot (with error bars) of μd 1 (3 objects) of Cluster 1

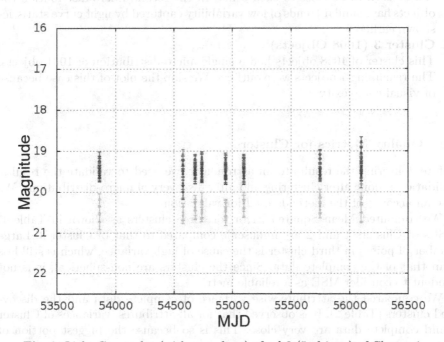

Fig. 6. Light Curve plot (with error bars) of μd 2 (5 objects) of Cluster 1

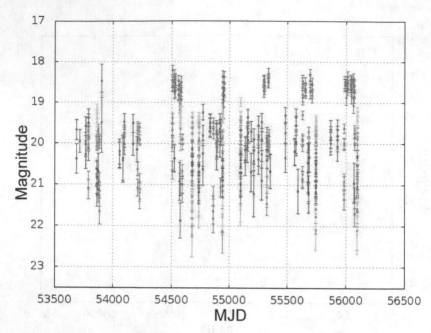

Fig. 7. Light Curve plot (with error bars) of μd 3 (11 objects) of Cluster 1

can be seen to be close to each other. In the second micro-distribution the objects have similar trends of low variability captured by light curve statistics shown earlier.

iii. **Cluster 3 (1008 Objects)**

This cluster of 1008 objects had a single micro-distribution of 1001 objects. The remaining 7 objects were outliers. We skip the plot of this case because of visual complexity.

3.4   Quality Metrics for Clusters

Before delivering the results to an astronomer, we need to validate the results. Validation is done after clustering and after discovery of micro-distributions. We give an account of the methods used for validation.

We computed Mean Squared Error for three clusters as shown in Table 1. MSEs of Clusters 1 and 2 are minuscule compared to that of Cluster 3. Large number of points in third cluster is the cause of high variance, which is still less than that of the complete data. Since the clusters are non-spherical, it is not prudent to consider MSE as a reliable metric.

We compared the attribute-wise variance of complete data and the discovered clusters (Table 2). It is observed that for all attributes, variances of Cluster 3 and complete data are very close. This is so because the biggest portion of

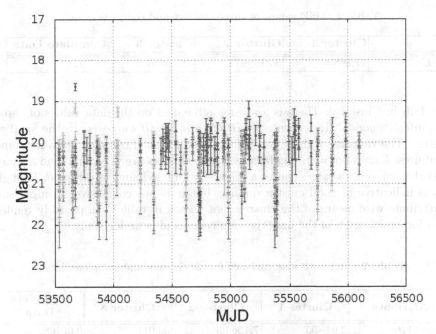

Fig. 8. Light Curve plot (with error bars) of μd 1 (9 objects) of Cluster 2

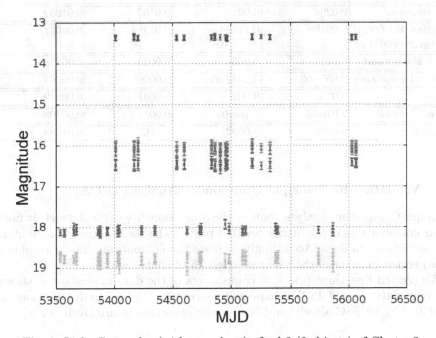

Fig. 9. Light Curve plot (with error bars) of μd 2 (6 objects) of Cluster 2

Table 1. MSE values of three clusters and complete data

	Cluster 1	Cluster 2	Cluster 3	Complete Data
MSE	11.86804	5.315997	419.6893	717.924

data falls in Cluster 3. There is a strong agreement on the data values of three attributes (*beyondstd*1, *percent_amplitude* and *small_kurtosis*) in the smaller clusters. The variance of these two is larger than that of complete data for some attributes. This is not alarming because these clusters are very small and a small perturbation in data values causes a relatively larger change in variance. Though this is indicative of good cluster quality, but it is not conclusive. Comparison of attribute-wise means of the cluster (not shown in table) can possibly render clues for explanation of the underlying physics and is under study.

Table 2. Attribute-wise variance for all clusters and complete data

Attribute	Cluster 1	Cluster 2	Cluster 3	Complete Data
beyond1std	2.04e-05	7.15e-05	0.0011	0.00108
fpr20	0.0202	0.0138	0.02197	0.02198
fpra50	0.0971	0.02995	0.0527	0.05331
linear_trend	0.0204	0.0029	0.0192	0.01902
median_buffer_range_percentage	0.0839	0.11394	0.07989	0.08067
pair_slope_trend	0.0135	0.0056	0.0083	0.00835
percent_amplitude	2.657e-05	4.243e-05	0.00066	0.00064
skew	0.0152	0.002	0.0081	0.00819
small_kurtosis	2.04e-05	7.15e-05	0.0011	0.00108
magratio	0.0202	0.0092	0.0169	0.01688

3.5   Validation by Principal Component Analysis (PCA)

Principal Component Analysis reduces the dimensionality of the dataset. It finds linear combinations of the attributes that prove to be the strongest contributors to the variance in data. Accordingly, it gives the components (linear combinations) sorted in order of their scores (strength of contribution).

We plotted first three principal components of the data in clusters as shown in Figures 10, 11 and 12. We expected to visualize the micro-distributions as identified by DUSSC algorithm. Cluster-wise observations are given below.

Cluster 1: Figure 10 shows 24 objects of this cluster sparsely spread over the 3-d plot. Five points in red on top left of the cube and three black points on top right are grouped together. Eleven points of this cluster however are *relatively* far, though grouped cohesively in the complete set.

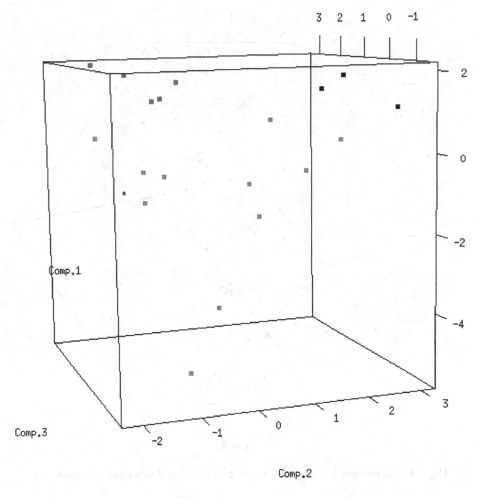

Fig. 10. Micro-distributions in Cluster 1 (plot using 3 principal components)

Cluster 2: Like Cluster 1, Cluster 2 is apparently sparse. Two micro-distributions are clearly visible in 2 different planes in Figure 11.

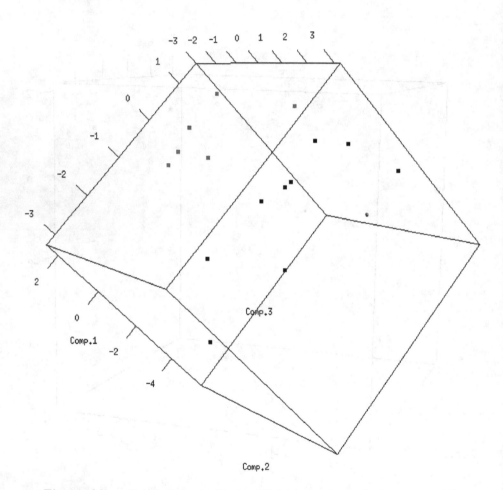

Fig. 11. Micro-distributions in Cluster 2 (plot using 3 principal components)

Cluster 3: Among the three clusters, Cluster 3 is the most cohesive because it houses more than 99% of the clustered data. In Figure 12, most of the objects of the cluster are in close proximity, resulting in the cluster's dense nature. In this dense cluster a single micro-distribution is evident.

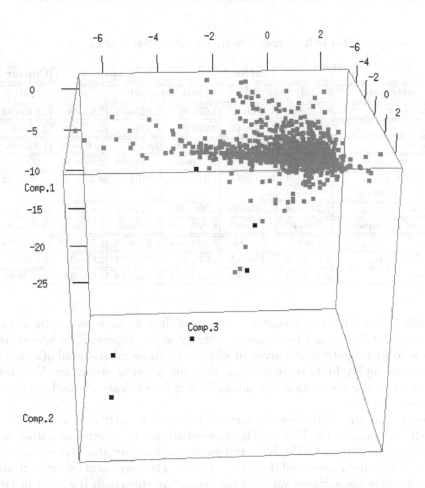

Fig. 12. Micro-distributions in Cluster 3 (plot using 3 principal components)

Thus, from each of the cluster's PCA plots, it is evident that there are certain smaller groups of objects contained in the clusters, which are closer to each other as compared to other objects and are presumably more similar in characteristics. Six smaller and more cohesive clusters i.e. micro-distributions have been correctly identified by the algorithm.

3.6   Validation by Statistical Methods

Variance is a statistical method to measure average deviation from mean. A smaller variance value indicates the lower variability of the data points and thus signifies more similarity among them. Therefore, another means by which we attempted to validate the micro-distributions captured by the algorithm, was with comparative analysis of attribute-wise variances. Table 2 and Table 3 enlist the variance values of the clusters and the micro-distributions, respectively.

Table 3. Variance values for micro-distributions

Attribute	Cluster 1			Cluster 2		Cluster 3
	μd_1	μd_2	μd_3	μd_1	μd_2	μd_1
beyond1std	6.653e-08	1.476e-07	2.67e-05	1.023e-04	2.63e-08	2.843e-04
fpr20	2.925e-03	1.096e-02	0.01945	1.968e-02	7.075e-03	2.154e-02
fpra50	0.0025	0.02493	6.964e-02	0.0389	1.067e-02	5.237e-02
linear_trend	3.433e-03	0.00122	3.742e-03	3.744e-03	2.567e-04	1.901e-02
median_buffer_range_percentage	6.333e-04	0.00195	1.31e-02	0.0164	9.467e-04	7.989e-02
pair_slope_trend	7.744e-04	3.005e-03	1.298e-02	5.734e-02	4.611e-03	8.106e-03
percent_amplitude	1.791e-07	2.449e-07	3.922e-05	5.491e-05	5.711e-08	3.087e-04
skew	2.15e-03	2.24e-03	2.2003e-03	1.337e-03	1.754e-03	8.167e-03
small_kurtosis	6.653e-08	1.476e-07	2.67e-05	1.023e-04	2.63e-08	2.843e-04
magratio	3.703e-04	2.056e-02	4.849e-03	3.333e-03	8.63e-03	1.604e-02

Table 3 shows that μd_1 of cluster 1 and μd_2 of cluster 2 have lower variances for almost all attributes and hence are apparently most cohesive. If an astronomer choose to investigate light curves of objects in these micro-distributions, the objects are highly likely to share some common physical properties. Variances in other $\mu d s$ are also reasonably small. This is an evidence of good quality of clusters.

Figure 13 to Figure 16 show the variance plots of four attributes selected based on MIC values shown in Table 4. The selected attributes *percent_amplitude* and *skew* have the highest MIC values, and *magratio* and *pair_slope_trend* have the lowest MIC values, among all the ten attributes. The horizontal line with cluster label indicates the variance values of the respective clusters. If the point plotted for the micro-distribution lies below its parent cluster's horizontal line, then the micro-distribution has a lower variance and is more cohesive compared to the parent cluster. Thus, with respect to the selected attributes, it is clearly evident that the cohesion in most of the micro-distributions is more than their respective parent clusters. Statistically, we have therefore substantiated the fact that the more homogeneous patterns present in the clusters have been captured in the form of six micro-distributions by the algorithm.

Table 4. MIC values for the attributes

Attribute	MIC Values
beyond1std	0.06824
fpr20	0.03497
fpra50	0.06137
linear_trend	0.0267
median_buffer_ range_percentage	0.05798
pair_slope_trend	0.01978
percent_amplitude	0.09102
skew	0.08551
small_kurtosis	0.06824
magratio	0.01542

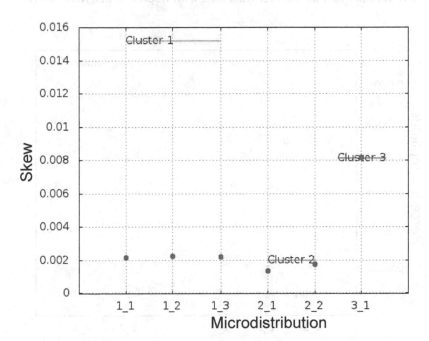

Fig. 13. Variance plot for microdistributions w.r.t. Skew Attribute

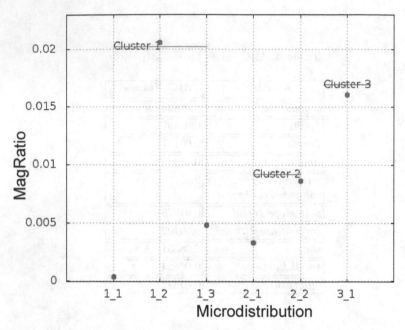

Fig. 14. Variance plot for microdistributions with respect to MagRatio Attribute

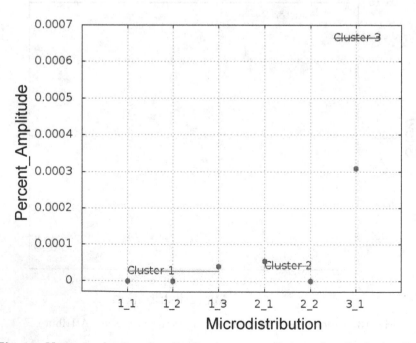

Fig. 15. Variance plot for microdistributions w.r.t. Percent_Amplitude Attribute

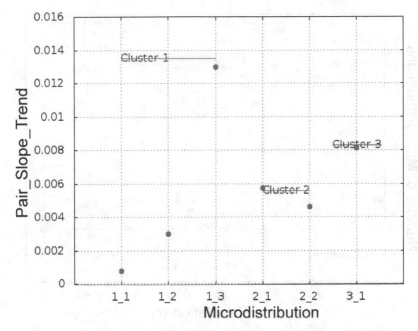

Fig. 16. Variance plot for microdistributions w.r.t. Pair_Slope_Trend Attribute

3.7   How to Determine Resolution?

Determining the right resolution for analyzing data to discover micro-distributions is crucial for success of the endeavor. Crafting an automated method for determining the best resolution for study is beyond the scope of this study. We followed trial-and-error strategy for choosing the *best* resolution. The heuristics used is described in detail below. We plotted the number of micro-distributions obtained at different resolutions as shown in Figure 17. The following observations are evident from the graph.

i. **Cluster 1 (24 objects)**: For resolutions upto 0.25, there is not much disintegration. All the data points are located in different cells and hence do not cluster. As r increases, number of micro distributions increases to a max of 3 micro-distributions at 0.45. The fact that the lower values of r do not result into micro-distributions indicates that cluster 1 is sparse. Sparse distribution of the objects belonging to this cluster is validated by the PCA plot of cluster 1 (Figure 10).

ii. **Cluster 2 (15 objects)**: Behaviour similar to cluster 1 is observed here too. At $r = 0.45$, two micro-distributions are discovered. Sparseness of the cluster is also validated by the PCA plot of cluster 2 (Figure 11).

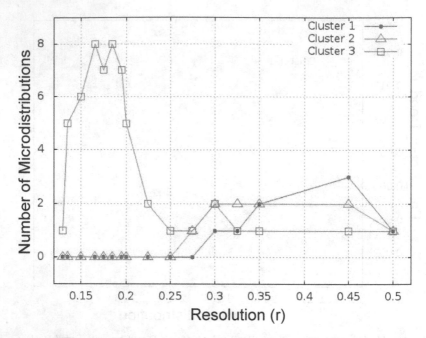

Fig. 17. Plot showing r v/s number of micro-distributions

iii. **Cluster 3 (1008 objects):** For lower resolutions the cluster reveals its constitution of up to 8 micro-distributions. For $r > 0.3$, only one micro-distribution with 1001 points was observed. Deeper analysis divulged that the micro-distribution accounted for 56.71% of total data. Cluster data other than this micro-distribution were distributed sparsely spread around the core and were branded as outliers. With increasing values of r (coarse resolution), the number of discovered micro-distributions falls sharply. Structure of the cluster with small dense core and larger spread is clearly visible in the PCA plot of Cluster 3 (Figure 12).

Equipped with these observations, we chose 0.45 as the best resolution for this data. Smaller clusters and fewer objects (Cluster 1 and 2) are more interesting and suitable for study by human experts compared to a large cluster. Such explorations aid data reduction and hence are instrumental for understanding of data by human experts.

4   Scalable Exploratory Analysis

Currently we have processed toy data-set of less than 2K objects. CRTS alone has 500 million light curves waiting to be investigated. With LSST survey opening up in few years, astronomers are dashing to develop techniques to analyze petabytes of data. This is indeed a case of *Big Data* characterized by *volume*, *variability, velocity* and *veracity*. It is imperative to ensure scalability of methods and algorithms developed for Astroanalytics for future. One of the common approaches to ensure this scalability is the development of parallel and distributed data mining algorithms. Development of incremental algorithms is another way to achieve the same goal. In the rest of the section we present a scrutiny of scalability of the proposed exploratory method.

As mentioned earlier, amount of light curve data currently available is massive and it is further growing at an exponential pace. In this use-case of *Big Data Analytics*, development of incremental and data parallel solutions is inevitable. Fortunately, the work-flow shown in Figure 4 is amenable to data parallel implementation on Hadoop Map-Reduce[2]. A Map-Reduce algorithm is a sequence of jobs, each of which runs on a subset of data in parallel fashion on multiple compute nodes. The underlying distributed file system (Hadoop DFS[27]) supports parallelism by replicating data on multiple data nodes.

Step A in the work-flow cleans and transforms the light curves. Since treatment of each light curve is independent of the other, this task is trivially amenable to data parallelism. It requires only one Map-Reduce job for cleaning as well as for computing statistics to form curated data set. Data set is *chunked* by HDFS, and each *chunk* is processed by a Map task. A Map instance checks the validity of the observation (record) and sends it to the Map-Reduce framework. A Reduce instance accumulates all the records with same object id to compute desired statistics and writes on HDFS. Hence, the time taken by this step is reduced by a factor of number of machines that are present in a cluster.

Step B is the clustering step, which takes curated data from HDFS as input. The algorithm maps each data point to an appropriate cell in 10-dimensional space. Each cell has a unique signature corresponding to its location. Since assignment of each data point to cell is independent of other, this processing can be done in parallel. Map task accomplishes this step. Reduce task accumulates count of data points in a cell. The algorithm proceeds to form clusters from these cells using *Connected Component Analysis* algorithm in parallel fashion. Making the algorithm incremental requires preserving the synopsis of data for use in the next iteration. The new points, in next iteration are merged with the saved synopsis. This scalable incremental clustering algorithm has been implemented and tested.

Since only significant clusters are selected for furher analysis, subsequent steps handle much less data. Labelling the clusters in Step C requires accessing the cluster members from the original data set, which requires only one MR job. Step D deals with small amount of data, since the size of clusters obtained is much smaller than the original data. Therefore, application of DUSSC algorithm for discovering micro-distributions can be done in sequential manner.

All MR algorithms implemented for the workflow fall in MRC^0 class because, (i) the algorithm takes limited number of rounds for completion, (ii) number of machines is sublinear to the size of the data and (iii) size of the data generated by every map and reduce instance is sublinear to the size of the data. Explanation

---

[2] Hadoop Map-Reduce is a open-source implementation of Google Map-Reduce[24].

of this model of complexity is beyond the scope of this study and interested reader is referred to [32].

Clustering component of the scalable work-flow has been implemented and tested. It is able to cluster 10-dimensional synthetic data set containing 0.4 billion points in approximately 20 minutes [38]. Testing the complete work-flow for *big data* is on agenda for future work.

5   Conclusion

Goals of the sky surveys present unprecedented challenges and opportunities for computer science research by way of posing fantastic demands for infrastructure as well as analytics. Astroinformatics is a multi-disciplinary field of research to solve astronomy and astrophysics problems in the era of digital astronomy, enabled by advances in computer science and engineering. It in turn enables astroanalytics that encompasses application of data mining, machine learning and statistical techniques on astronomy data for making sense out of colossal data sets acquired by sky surveys.

We presented the challenges and opportunities to which analytics research has been exposed by astronomy data deluge. As a case study, we presented a workflow to segregate light curves using exploratory data analysis. We use a tweaked version of an existing algorithm to discover clusters and then perform deep analysis to investigate heterogeneity in the clusters. The obtained results were checked for quality using statistical methods and were found to be satisfactory. Scientific analysis and interpretation of the results is pending. We have tested the scalability of the clustering component and testing the complete work-flow is the agenda for future work.

Acknowledgement. We gratefully acknowledge the help and support rendered by Ashish Mahabal during the course of this work. We thank him for passing on to us the domain knowledge, giving directions and suggestions that were instrumental in shaping the analysis.

References

1. http://www.ivoa.net/
2. http://www.sdss.org/
3. http://crts.caltech.edu/
4. http://ptf.caltech.edu/iptf/
5. http://www.sciops.esa.int/index.php?project=ASTROF&page=index
6. http://wise.ssl.berkeley.edu/astronomers.html
7. http://www.lsst.org/lsst/
8. https://www.skatelescope.org/
9. http://www.astro.princeton.edu/PBOOK/datasys/datasys.htm
10. http://www.startap.net/starlight/
11. http://boinc.berkeley.edu/
12. http://avyakta.caltech.edu/science/datasets/SAMSI_DC/index.html

13. http://nirgun.caltech.edu:8000/scripts/description.html
14. Mahabal, A., Djorgovski, S., Drake, A., Donalek, C., et al.: Discovery, classification, and scientific exploration of transient events from the Catalina Real-time Transient Survey, arXiv:1111.0313v1
15. Aggarwal, C.C. (ed.): Data Streams - Models and Algorithms. Advances in Database Systems, vol. 31. Springer (2007)
16. Agrawal, R., Gehrke, J., Gunopulos, D., Raghavan, P.: Automatic subspace clustering of high dimensional data for data mining applications. SIGMOD Rec. 27(2), 94–105 (1998)
17. Szalay, A.S., Kunszt, P., Thakar, A., Gray, J., Slutz, D.: The Sloan Digital Sky Survey and its Archive, arXiv:astro-ph/9912382v1
18. Ball, N.M., Brunner, R.J.: Data mining and machine learning in astronomy. International Journal of Modern Physics D 19(07), 1049–1106 (2010)
19. Bhaduri, K., Das, K., Borne, K.D., Giannella, C., Mahule, T., Kargupta, H.: Scalable, asynchronous, distributed eigen monitoring of astronomy data streams. Statistical Analysis and Data Mining 4(3), 336–352 (2011)
20. Bhatnagar, V., Dobariyal, R., Jain, P., Mahabal, A.: Data understanding using semi-supervised clustering. In: CIDU, pp. 118–123. IEEE (2012)
21. Bhatnagar, V., Kaur, S., Chakravarthy, S.: Clustering data streams using grid-based synopsis. Knowledge and Information Systems, 1–26 (2013)
22. Chen, Y., Alspaugh, S., Katz, R.: Interactive analytical processing in big data systems: A cross-industry study of mapreduce workloads. Proc. VLDB Endow. 5(12) (August 2012)
23. de Andrade Silva, J., Faria, E.R., Barros, R.C., Hruschka, E.R., de Carvalho, A.C.P.L.F., Gama, J.: Data stream clustering: A survey. ACM Comput. Surv. 46(1), 13 (2013)
24. Dean, J., Ghemawat, S.: Mapreduce: Simplified data processing on large clusters. In: Proceedings of the 6th Symposium on Operating System Design and Implementation, pp. 137–150 (2004)
25. Domingos, P., Hulten, G.: Mining high-speed data streams. In: Proceedings of the Sixth ACM SIGKDD International Conference on Knowledge Discovery and Data Mining, pp. 71–80. ACM, New York (2000)
26. Dutta, H., Giannella, C., Borne, K.D., Kargupta, H.: Distributed top-k outlier detection from astronomy catalogs using the demac system. In: SDM (2007)
27. The Apache Software Foundation. Welcome to Hadoop™; Distributed File System (2007)
28. Han, J., Kamber, M., Pei, J.: Data Mining: Concepts and Techniques, 3rd edn. Morgan Kaufmann Publishers Inc., San Francisco (2011)
29. Heer, J., Kandel, S.: Interactive analysis of big data. XRDS 19(1), 50–54 (2012)
30. Heer, J., Shneiderman, B.: Interactive dynamics for visual analysis. Commun. ACM 55(4), 45–54 (2012)
31. Mambretti, M.B.J., DeFanti, T.: Starlight: Next-generation communication services, exchanges, and global facilities (chapter). Advances in Computer 80, 191–207 (2010)
32. Karloff, H., Suri, S., Vassilvitskii, S.: A model of computation for mapreduce. In: Proceedings of the Twenty-First Annual ACM-SIAM Symposium on Discrete Algorithms, SODA 2010, pp. 938–948. Society for Industrial and Applied Mathematics, Philadelphia (2010)
33. Keim, D.A.: Visual exploration of large data sets. Commun. ACM 44(8), 38–44 (2001)

34. Borne, K.D.: Astroinformatics: A 21st Century Approach to Astronomy, arXiv:0909.3892v1
35. Borne, K.D.: Scientific Data Mining in Astronomy, arXiv:0911.0505v1
36. Nigro, S.E.G.C., Oscar, H., Xodo, D.H.: Data Mining with Ontologies: Implementations, Findings, and Frameworks. IGI Global (2008)
37. Lupton, R., Gunn, J.E., Ivezic, Z., Knapp, G.R., Kent, S., Yasuda, N.: The SDSS Imaging Pipelines, arXiv:astro-ph/0101420v2
38. Kaur, S., Saxena, R., Khanna, D., Bhatnagar, V.: Comparing data processing frameworks for scalable clustering. To appear in Proceedings of FLAIRS 2014, to be held in (May 2016)
39. Simoff, S.J., Maher, M.L.: Ontology-based multimedia data mining for design information retrieval. In: Proceedings of ACSE Computing Congress, vol. 320, ACSE, Cambridge (1998)
40. Singh, S., Vajirkar, P., Lee, Y.: Context-based data mining using ontologies. In: Song, I.-Y., Liddle, S.W., Ling, T.-W., Scheuermann, P. (eds.) ER 2003. LNCS, vol. 2813, pp. 405–418. Springer, Heidelberg (2003)
41. Thompson, D., Burke-Spolaor, S., Deller, A., Majid, W., Palaniswamy, D., Tingay, S., Wagstaff, K., Wayth, R.: Real time adaptive event detection in astronomical data streams: Lessons from the very long baseline array. IEEE Intelligent Systems 99, 1 (2013)
42. York, D.G., et al.: The Sloan Digital Sky Survey: Technical Summary. Astron. J. 120, 1579–1587 (2000)
43. Zudilova-Seinstra, E., Adriaansen, T., van Liere, R.: Trends in Interactive Visualization: State-of-the-Art Survey, 1st edn. Springer Publishing Company, Incorporated (2008)

Implementing Agent-Based Resource Management in Tsunami Modeling – Preliminary Considerations

Michał Drozdowicz[1], Kensaku Hayashi[2], Maria Ganzha[1], Marcin Paprzycki[1], Alexander Vazhenin[2], and Yutaka Watanobe[2]

[1] Systems Research Institute Polish Academy of Sciences, Warsaw, Poland
name.surname@ibspan.waw.pl
[2] Graduate School Department, University of Aizu, Aizu-Wakamatsu, Japan
{vazhenin,yutaka,m5161111}@u-aizu.ac.jp

Abstract. Recently, work has started to apply the agent-semantic infrastructure, developed within the scope of the *Agents in Grid* project, to the resource management needed in tsunami modeling. The original proposal was based on the perceived simplicity, versatility and flexibility of the agent-based approach that makes it easier to deploy than the standard grid middlewares. The aim of this paper is to report on the progress in implementing and deploying the proposed system at the University of Aizu.

1   Introduction

One of the key effects of the Great Japanese Earthquake and Tsunami was restatement of the importance of studying the impact of such events at different time scales. Here, the two main issues that have to be addressed are: (1) real-time tsunami warning, and (2) long-term hazard assessment. To respond to these needs, it is necessary to: (a) facilitate use of distributed clients, providing access to computational services, (b) address scalability, to allow an arbitrary number of users and computational resources to interact in a customizable working environment, and (c) since different applications and services may be developed for a variety of hardware/software platforms, reusability and interoperability are important aspects of applying (and, possibly, combining) computational resources and services [1, 2].

Here, observe that personal computers (with, or without, additional enhancements, such as GPU processors) can be used for high-demand computing applications, e.g. consider the Folding@home project that involves distributed PC-based simulations of protein folding and other molecular dynamics simulations [3, 4]. Unfortunately, while most volunteer projects (e.g. projects based on the BOINC infrastructure) do note require human supervision – results are collected / accumulated during a long-term execution, and only in the final stage they are inspected by the humans, this is not the case with tsunami modeling. Here, the typical research scenario requires human interactions during the process,

A. Madaan, S. Kikuchi, and S. Bhalla (Eds.): DNIS 2014, LNCS 8381, pp. 95–111, 2014.
© Springer International Publishing Switzerland 2014

e.g. checking/understanding the results of the current test(s) is often required to instantiate the next round of experiments. Furthermore, different tsunami models often need to be combined to study various phases of the tsunami phenomenon. This makes a direct application of the BOINC-like approach difficult, if not impossible.

As a result, in [20] it was stipulated that an agent-semantic system, designed for resource management in the grid (developed in the *Agents in Grid* project; *AiG*), can be successfully applied in tsunami modeling research. The aim of this paper is to discuss how the *AiG* approach can be used to instantiate a distributed tsunami modeling laboratory. To this effect, we start with a brief summary of the state-of-the-art in tsunami modeling. Next, we outline the MOST algorithm designed for tsunami simulation, and provide details of the implementation developed and used at the University of Aizu. We follow with an overview of the Agents in Grid system and discuss its key aspects related to the process of its customization to the requirements of tsunami modeling. Finally, we provide a description of the process of specifying and submitting a job within the distributed tsunami modeling laboratory. This process is used to illustrate key features of the system under development.

2   Tsunami Modeling State-of-the-Art

Let us start from briefly discussing the state-of-the-art in tsunami modeling. In [5], authors suggested that complex mathematical models and high mesh resolution should only be used when necessary. They have developed a "parallel hybrid tsunami simulator," based on mixing different models, methods and meshes. This simulator was implemented using object-oriented techniques, allowing for easy reuse of existing codes. Here, high performance was not the main goal. Instead, research was focused on combining various approaches to develop high quality hybrid tsunami models.

Authors of [6], experimented with eight different parallel tsunami propagation simulators. Each of them used a mixed-mode programming model, consisting of a thread-based shared memory part, a distributed memory part and, finally, a virtual shared memory-based part. Obtained results have illustrated various problems with scalability of the investigated software artifacts. Furthermore, it was shown that if sufficient node memory is not available, threading becomes the bottleneck.

The TsunamiClaw is a software package based on a finite volume method [7]. It solves the shallow water equations in the, physically relevant, conservative form. Thus, the obtained solution is represented as water depth and momentum. Currently, this project is no longer actively pursued. Instead it has been generalized into the GeoClaw software.

The TUNAMI-N2 software [8], is a tsunami simulation, which uses separate models for the deep sea and shallow water. Interestingly, it uses constant grid size in the entire domain. The TUNAMI was originally authored by by Imamura (in 1993) and later applied to the real tsunami events in many countries. The package was written in FORTRAN and has a standard GUI.

Finally, the MOST (Method of Splitting Tsunami) software allows for real-time tsunami inundation forecasting, by incorporating real-time data from actual detection buoys [9,10]. Furthermore, in the US, the MOST model is used for developing inundation maps [11]. To use in computational practice, a web enabled interface, named ComMIT, has been implemented.

3 Tsunami Modeling Environment and Processes

3.1 Basic Model and Software Tools

As discussed, there exist multiple tsunami models and algorithms implemented to realize them. These models deal with: origins of tsunamigenic earthquakes (estimation of magnitude and epicenter location), determination of the initial displacement at the tsunami source, wave propagation, inundation into the dry land, etc. Overall, the tsunami modeling environment is typically used to simulate three phases of the tsunami evolution: (1) estimation of residual displacement area, resulting from an earthquake and causing the tsunami, (2) transoceanic propagation of the tsunami through the deep water, and (3) contact with the land (run-up and inundation). Out of available choices, the University of Aizu team selected the MOST package [9,10]. The MOST approach was initially developed in the Tsunami Laboratory of the Computing Center of the USSR Academy of Sciences in Novosibirsk. Subsequently, the method was updated in the National Center for Tsunami Research (NCTR, Seattle, USA) and adapted to the standards accepted by tsunami watch services in the US, as well as other countries. It was then used in tsunami research in many countries around the world. According to this approach, propagation of the wave in the ocean is governed by shallow-water differential equations:

$$H_t + (uH)_x + (vH)_y = 0,$$
$$u_t + uu_x + vu_y + gH_x = gD_x,$$
$$v_t + uv_x + vv_y + gH_y = gD_y, \tag{1}$$

where $H(x,y,t) = h(x,y,t) + D(x,y,t)$; h - is the water surface displacement, D - depth, $u(x,y,t)$ and $v(x,y,t)$ - are the velocity components along the x and y axis', g - is the gravity. The initial conditions should confirm the presence of water in all grid points, except for the tsunami source, where the surface displacement is not equal to zero.

The numerical algorithm splits the difference scheme, which approximates equations (1) in the spatial directions. A finite difference algorithm, based on the splitting method, reduces the solution of equations with two space variables to the solution of two one-dimensional equations. As a result, effective finite difference schemes, developed for the one-dimensional problems, can be applied. Moreover, this method permits to set boundary conditions for a finite-difference boundary value problem, using a characteristic line method.

3.2 General Calculation Process

Figure 1 shows the block-diagram illustrating the overall structure of calculations. To run the program, it is necessary to specify:

- bottom topography or bathymetry data;
- initial and boundary conditions;
- modeling parameters such as time-steps and length of the model run.

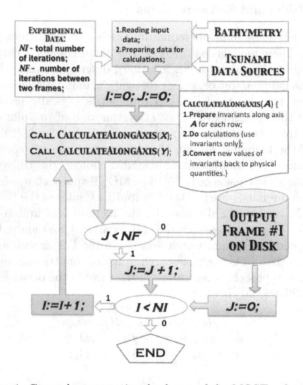

Fig. 1. General computational scheme of the MOST software

The necessary parameters are passed to the modeling program as a *scenario file* that is composed of the following elements:

- Area Information – BathymetryArea. Containing:
  - Grid Name – Name of the Sea Area under consideration
  - Grid Axes Version – Version Number
  - Grid File Name – Link/Path to the Bathymetry Data/File
- Computational Parameters – CalculationInformation. Containing:
  - Minimum Depth – Minimum depth for the offshore area
  - Time step – Time between two modeling iterations (in seconds)
  - Total number of steps – The total number of iterations

- Number of steps between snapshots – Number of output frames (NF)
- Modeling time passed – The time elapsed since the beginning of modeling
- Save output every n-th grid point – Specification of results saving structure
- Global b.c.s – 1=global, 0=non-reentrant
- Naming Rules – NameOfOutputResult. Containing:
  - Filename – <prefix_ha.nc>, or "auto"
  - Source Zone Name – Name of domain inside the bathymetry area
  - Source Zone Code – Code of domain
  - Source Column – Code of Grid (Column)
  - Source Row – Code of Grid (Row)
  - Source Version – Version of source
- Fault Plane Information – InformationAboutEarthquakeData. Containing:
  - Number of Fault Planes – Information concerning number of fault planes
  - x-integration – X value for integration
  - y-integration – Y value for integration
  - Vp – P-wave velocity
  - Vs – S-wave velocity
  - Deform Area X – Value of the Deform Area (x-axis)
  - Deform Area Y – Value of the Deform Area (y-axis)
  - Longitude (deg) – Center of the initial Wave (Longitude)
  - Latitude (deg) – Center of te initial Wave (Latitude)
  - Length (km) – Size of the initial Wave (Length)
  - Width (km) – Size of the initial Wave (Width)
  - DIP (deg) – DIP of the Wave
  - RAKE (deg) – RAKE of the Wave Form
  - STRIKE (deg) – Strike of the Wave
  - SLIP (m) – Slip of the Wave
  - DEPTH (km) – Depth of the Wave

The complete model of this information is depicted in Figure 2. As will be seen, this information was used to develop the initial version of tsunami modeling ontology.

After launching, the program implements calculations and stores results as a series of frames representing tsunami propagation process in time. Parameter NF defines the time interval, during which the results of computations are persisted in the computer memory. After this time expires, results are stored on the secondary storage devices in the NetCDF format ([11]).

It is important to note that it took about 3.31 seconds to complete a single time step of the original (Fortran 90) program on a computer with 4 dual-core, Intel Xeon 2.8GHz, CPUs. After the program was ported to C/C++, it takes about 3.00 seconds for a single time step [12]. Since a typical simulation, consists of about 10000 time steps, it requires about 8 hours to complete. Therefore, the tsunami modeling needs to be significantly accelerated (e.g. through parallel processing); especially for real-time tsunami warning generation. However, speeding up modeling is also crucial for repetitive tsunami simulations; e.g. in the artificial island modeling scenario.

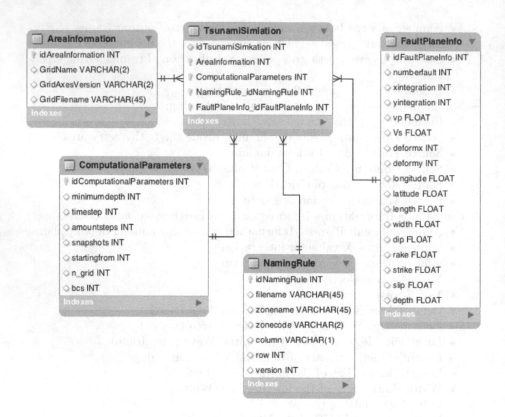

Fig. 2. Scenario parameters

3.3   Hybrid Tsunami Modeling Combining Natural and Artificial Bathymetry Objects

Lessons from the Great Japanese Tsunami stress importance of: (i) being able to provide real-time tsunami warning, (ii) long-term hazard assessment (e.g. running detailed inundation models across the Japanese sea-line, and (iii) studies of the well-known "Matsushima effect." The later concerns the influence of geographical objects, like islands, on the wave height and/or speed. This observation finds its foundations in the research reported in [14]. Here, results of an experimental study on effects of submarine barriers on tsunami wave propagation indicated capability of reducing tsunami run-up through strategic placement of artificial objects interacting with the tsunami waves. Specifically, it may be possible to design and build a set of artificial objects (islands) that can be used to protect the coastal areas. In particular, such protection could be of extreme value in highly populated areas (e.g. coastal cities, such as Sendai, that was affected by the tsunami of 2011), as well as in industrial areas (e.g. nuclear plants, factories, airports, etc.). Note that such effect, caused by local islands, already exist in the Matsushima area, while absent on the Fukushima coast. It can be conjectured that adding a small number of appropriately placed artificial objects "in front of" the Fukushima Nuclear Plant could have mitigated the effect of the tsunami and prevent the disaster. For more details, see also [19, 20].

4 AiG for Tsunami Modeling

Let us now discuss how the *Agents in Grid (AiG)* project can be used to support tsunami modeling. We will start from an overview of these parts of the *AiG* approach that are pertinent in the current context. The *AiG* project aims at providing a flexible agent-based infrastructure for managing resources in the grid ([15, 16]). Application of software agents and semantic technologies makes it well-suited for open, dynamic and heterogeneous environments. The *AiG* architecture is based on the premise of an open Grid – a network of heterogeneous resources, owned and managed by different organizations. It allows for users to either provide a new resource to the Grid in order to earn money, or to use the Grid to execute a task.

In the original *AiG* approach, each resource is governed by a *WorkerAgent* and performs its tasks as part of a team, managed by an *LMaster* agent. Teams are registered in a yellow-pages-like directory service, represented by the *Client Information Center* (*CIC*) agent, which handles the initial matchmaking of users to teams. Users interact with the system through their (dedicated) *UserAgents*. The decision, which team to choose to execute the job is a result of autonomous negotiations between the *UserAgent* and the *LMasters* of appropriate teams. In a similar way, adding a resource to the system involves negotiations with teams looking for new members. The main features of the system are shown in Figure 3, in the form of a use case diagram.

An important aspect of the project is the fact that all data and information in the system are represented in ontological format, using the OWL language. The usage of ontologies enables to describe jobs, resources and their relationships in a structured, yet flexible way (for more details, see [17]). Thanks to application of ontologies, providing support for new (added to the system) types of hardware and augmented software configuration (such as new software libraries or programs) involves only modification of the ontology terms and does not require any additional customization. As the job descriptions are also defined using the OWL language, it is possible to specify different parameter sets and requirements towards computing resources depending on the type of the task to be performed. For instance, as part of the initial implementation of the AiG system at the University of Aizu, we have extended our core ontology to be able to accurately describe the hardware configuration of the machines used in the experiments, as well as terms related to the tsunami modeling (see, Section 3.2).

Putting the AiG system to work for tsunami modeling we have made a few observations. First, computers made available to the tsunami research at the University of Aizu do not constitute an open environment where resources join and leave "dynamically". Second, there is no economic aspect – we can safely assume that if a resource matches the requirements of the job and is available

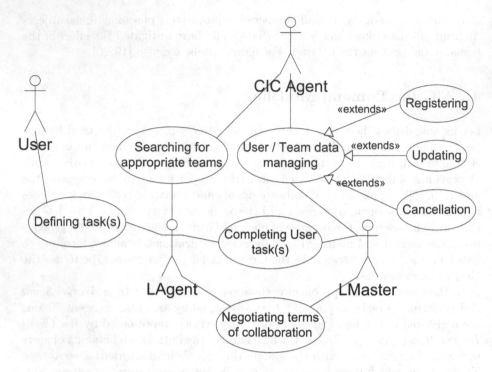

Fig. 3. Use Case diagram of the AiG system

for use, there is no need for negotiations concerning its price. However, what is still required are negotiations concerning availability of resources (e.g. when a given laboratory is in use during certain class periods and machines cannot be used for other purposes then instruction). To address these points we have modified the *AiG* system. First, we have resigned from the notion of resource team, and placed a *WorkerAgent* (playing the role of the *LMaster*) on each computing node. We have also eliminated the scenario in which the *WorkerAgent* is joining a team. Instead, adding a new resource means registering it with the *CIC Agent* as a standalone node (one-member team). Finally, in the next phase of the implementation, we will re-focus the negotiations. Their role will be to provide information about current and planned utilization of the resources. This will allow the *UserAgent* to decide where to run, which job, and when.

4.1   Job Submission Process – Summary

Let us now go through the entire process of conducting an experiment, using resources at the University of Aizu and use it to illustrate the details of the modified *AiG* system.

The user starts by accessing a web based interface, which is the entry point to the communication with the *UserAgent*. The next step is to specify the hardware requirements for the job in the form of constraints on the ontological terms

Fig. 4. Example of choosing a property for constraining

describing the resources. This task is done using the interface based on the On-toPlay module [18] (its Condition Builder component), giving the user complete freedom in describing the needed resources, while guiding her through the contents of the ontology without the need for deep knowledge of its structure (knowledge of semantic technologies, in general). As shown in Figure 4, the Condition Builder is composed of a series of condition boxes used to create constraints on class-property relationships. Depending on the chosen class, the user can select, which class property she wishes to restrict. For example, having selected the *GPUMemory* class, the expanded property box will contain properties such as *hasTotalSize* and *hasAvailableSize* (see, Figure 4).

After selecting the class and property the user can choose the required operator and value. Here, she sees only the operators applicable to the given type of the property. Specifically, this means that for value properties (such as amount of available GPU memory) it would be operators such as *equalTo*, *lessThan* or *greaterThan*, while for object properties (e.g. the CPU installed on the node) the user would be allowed to select, e.g. is equal to individual or is constrained by. Note that the selection of available operators is performed by the front-end, on the basis of the ontology and was *not* hand-coded. Should a user wish to restrict the value of a particular property to a fixed individual from the ontology, the Condition Builder lists all available individuals that can be used in the context (see, Figure 5).

Let us now assume that an object property is selected choosing the "is constrained by" operator. This enables the user to specify the type of object for which the value should be constrained, and to create additional constraints on that class (see, Figure 6).

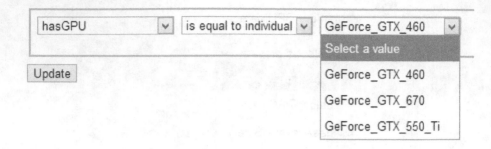

Fig. 5. Example of choosing an individual

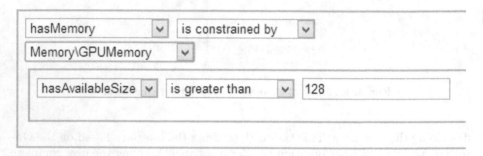

Fig. 6. Example of choosing a nested condition

4.2 Job Submission Process – Detailed Example

Let us now look at a practical example that will illustrate the entire process. As described in [9, 10, 12, 13], the implementation of the MOST code, used at the University of Aizu, is most effective when run on a CUDA-based GPU, with a sufficient amount of available GPU memory. Therefore, let us assume that the user wishes to schedule a job on a resource that has a GPU with at least 512 MB of available GPU memory (which is one of the machines available for our experiments). In this case the user starts with an empty *ComputingComponent* specification. First, she would constrain the property *hasGPU* of the class *ComputingComponent* to contain a value of type *GPU_CUDA*. Let us assume that it does not need to be any particular GPU model, so we do not add additional conditions on this class (though such specification already exists in the completed representation of available machines). Second, the user adds a condition on the property *hasMemory*, constraining it to the *GPUMemory* subclass and adding a nested condition specifying that the *hasAvailableSize* property should have a value greater than 512 MB. Figure 7 represents the completed condition. Once more, note that during the selection process, only these properties and individuals are "shown to the user" that have been specified within the ontology; and that no hand-coding of these terms and conditions was required. All "work" is done by the front-end on the basis of the ontology.

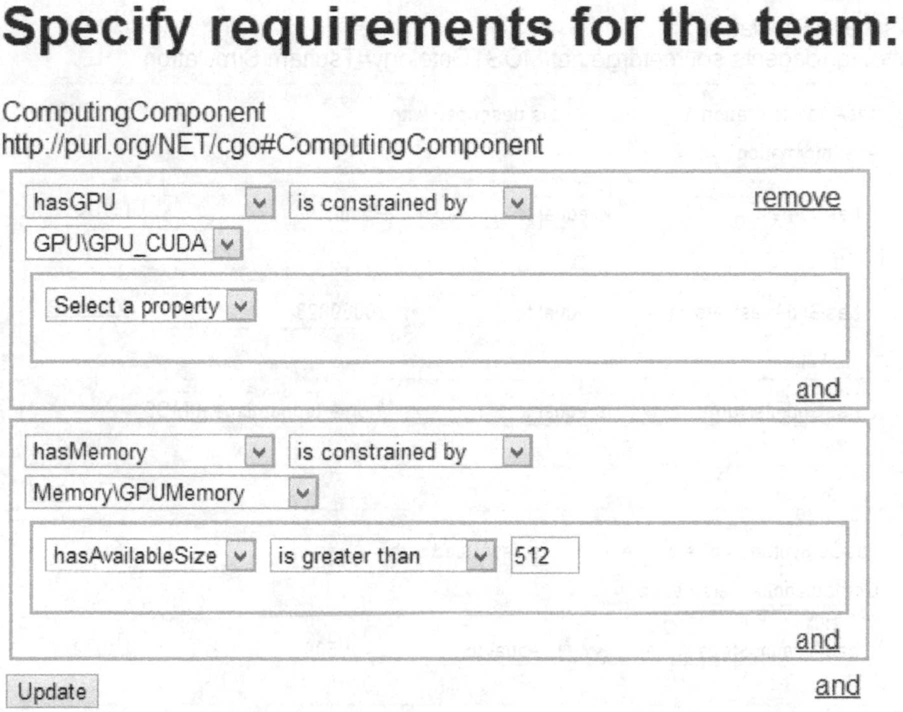

Specify requirements for the team:

ComputingComponent
http://purl.org/NET/cgo#ComputingComponent

hasGPU ▾	is constrained by ▾	remove
GPU\GPU_CUDA ▾		
Select a property ▾		

and

hasMemory ▾	is constrained by ▾	
Memory\GPUMemory ▾		
hasAvailableSize ▾	is greater than ▾	512

and

Update   and

Fig. 7. Complete requirements specification

After the user submits the resource requirements, the *UserAgent* passes this description to the *CIC Agent*, which performs semantic reasoning on its knowledge base, to find resources satisfying the given criteria and returns a list of matching nodes, including the information on how to contact the *LMasters* (in our case the *WorkerAgents*) at each node.

Note that in a dynamic environment, such as the university laboratory, there is no guarantee that the resources found by the *CIC Agent* are, at the moment, available for use. The machine might be offline, used for other purposes, or the agent process (and/or container) might not be running. Therefore, there is a need for additional verification of the availability of the resources. This is handled using the mechanism of multi-agent negotiations, albeit in a very simplified form. When the *UserAgents* receives the list of *LMaster* addresses, it issues a *Call For Proposal (CFP)* message to gain confirmation of whether the resources are able to perform the task. The *LMasters* confirm that this is the case (or reject the proposal), and provide information when they could start executing the job. This helps to handle the case of temporarily occupied nodes. Once the *UserAgent* receives offers from the agents (here, note that we assume their benevolence), it presents the list to the user, who can choose the node(s) on the basis of their availability and other parameters. Here, the resource selection may be also passed to the *UserAgent*, but this will require further considerations and will be approached in the next phase of the project.

TsunamiSimulation
http://gridagents.sourceforge.net/MOSTOntology#TsunamiSimulation

hasArialInformation ⌄	is described with ⌄		remove
ArialInformation ⌄			

hasName ⌄	is equal to ⌄	Pacific	remove

hasGridAxesVersion ⌄	is equal to ⌄	20060823	remove

hasGridFileName ⌄	is equal to ⌄	/home/tsunamiagent/MOS	remove

and

hasComputationalParameters ⌄	is described with ⌄	
ComputationalParameters ⌄		

hasAmountSteps ⌄	is equal to ⌄	500	remove

hasSnapshots ⌄	is equal to ⌄	4	remove

hasSaveOutputFrequency ⌄	is equal to ⌄	1	remove

hasTimeStep ⌄	is equal to ⌄	10	remove

hasMinimumDepth ⌄	is equal to ⌄	10	remove

hasStartingFrom ⌄	is equal to ⌄	6

and

Update   and

Fig. 8. Simulation scenario

4.3 Specifying the Scenario Description

The final step of submitting jobs is the specification of the executable code / library an of the necessary parameters. As described in the previous sections, for the tsunami simulations it is crucial to be able to run different kinds of algorithms on different data sets and variables to come up with collections of results (particularly in the case of tsunami modeling, rather than generating tsunami warnings). Consequently, in this case, the user is going to provide multiple job descriptions (one for each model / parameter set in the simulation). The job description is provided using *the same* Condition Builder mechanism, although using a different ontology.

As part of the implementation of the *AiG* system at the University of Aizu, a new ontology – the *MOSTOntology* – has been created to represent the entities forming the simulation scenario (recall that these parameters have been described in Section 3.2). This ontology is an extension of the *AigGridOntology* in a way that a newly introduced class `TsunamiSimulation` is introduced, which is a sub-class of the `JobDescription` class. Other classes contained in the *MOSTOntology* correspond directly to the entities from the scenario file:

- `AreaInformation`
- `ComputationalParameters`
- `NamingRule`
- `FaultPlaneInfo`

The ontology also contains object properties linking `TsunamiSimulation` with the above mentioned classes, as well as all data properties describing them (as specified in Section 3.2).

The introduction of the *MOSTOntology* into the *AigGridOntology* enables the user to specify the job using the same Condition Builder interface. It also makes it possible for the *WorkerAgent* to generate the scenario file from the ontological information, thus removing the need to deploy the scenario files onto each (potential) grid node. Of course, the system will still support running jobs using scenario files accessible locally on the nodes, or at a network location accessible to them, but the goal of achieving simplified access of users to distributed, heterogeneous resources has been achieved.

To give an example of how the contents of the scenario file corresponds to the contents of the *MOSTOntology* and to illustrate the usage of the Condition Builder for the job specification, Listing 1.1 contains a sample scenario file, while Fig. 8 depicts a part of the same scenario represented in the *AiG* user interface.

Listing 1.1. Sample scenario file

```
#   MOST Propagation test input file
#
#   This is the format for an input file for running the
#   MOST Propagation program version 1.3
#
#   Comments are prefixed with a hash "#", and can appear
#   on their own line, or after a parameter.
#    The only important thing is order of parameters
#
```

```
#  If there are multiple fault−planes, user must provide
#  all deformation parameters repeated for each fault−plane
#  (repeat from "x−integration" to "Depth" for each fault−plane)
#
#  If number of fault−planes is 0, MOST expects to read the
#  deformation from a file in the MOST grid (ASCII Grid) format
#  If number of fault−planes is < 0, MOST reads deform.dat file
#  created from the previous run of MOST (hint, use this to
#  keep from re−running deformation)
#
#  Beginning of test input file:
#        Grid Name
Pacific
#        Grid Axes Version
20060823
#        Grid Filename
/home/tsunamiagent/MOST/Bathymetry/pacific_4m_nocaribbean.corr
#        Computational parameters
20   # Input minimum depth for offshore (m)
10   # Input time step (sec)
1000    # Input amount of steps
6    # Input number of steps between snapshots
6    # ... Starting from timestep
4        # Save output every n−th grid point
0    # Input global b.c.s (1=global, 0=non−reentrant)
#        Output filename (<prefix>_ha.nc, or "auto")
auto
#        Source naming info
#   Source Zone Name
Aleutian−Cascadia
#    Source Zone Code (two characters)
ac
#   Source Column (one character)
b
#   Source Row (integer)
13
#   Source Version (integer)
0
#        Fault plane info
1            # Number of fault−planes
41           # x−integration
21           # y−integration
8.11         # Vp − P−wave velocity
4.49         # Vs − S−wave velocity
200          # Deform Area X
200          # Deform Area Y
179.842      # Longitude (deg)
51.085       # Latitude (deg)
100.0        # Length (km)
50.0         # Width (km)
15.0         # DIP (deg)
90.0         # RAKE (deg)
271.0        #STRIKE (deg)
1.0          # SLIP (m)
5.0          # DEPTH (km)
```

4.4 Job Execution

Once the user completes the job description, it is sent by the *UserAgent* to the respective *LMaster* (the one that was selected as a result of the, above described, negotiations), which then starts task execution. The information that is passed from the *UserAgent* to the *LMaster* is the ontology fragment, containing information needed to generate the scenario for the MOST software. When the

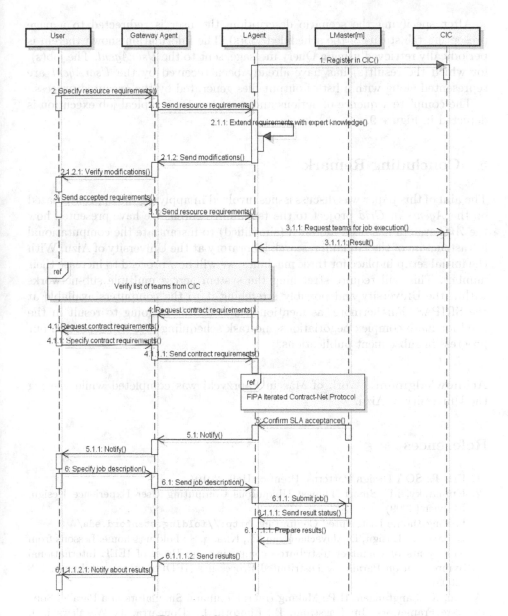

Fig. 9. AiG job execution sequence

computation is finished, the *LMaster* creates a JobResult message, which contains information about the job execution, the outcome and links to the result data and (any resources created by the simulation algorithm). The *UserAgent*, on the other hand, is responsible for gathering all responses from the nodes taking part in the experiment (in the case, when multiple simulations have been submitted to multiple nodes).

After specifying the scenario description, the user is redirected to a page presenting the status of the scheduled job(s). The information shown therein is periodically retrieved using a Query message, sent to the *UserAgent*. The job(s), for which the result(s) has/have already been received by the *UserAgent* are represented along with a list of output files generated by the executed process.

The complete sequence of actions and messages for a typical job execution is depicted in Figure 9.

5   Concluding Remarks

The aim of this paper was discuss issues involved in applying the approach based on the *Agents in Grid* project to the tsunami research. We have presented how the *AiG* system has been modified (simplified) to instantiate the computational infrastructure of the tsunami research laboratory at the University of Aizu. With the initial setup in place for three machines, we will now proceed to increase their number. This will require stretching the system across multiple sub-networks within the University and possibly stretching it to the computers available at the SRIPAS. Furthermore, as mentioned above, this is going to result in the need for more complex negotiations and task scheduling. We will report on our progress in subsequent publications.

Acknowledgment. Work of Marcin Paprzycki was completed while visiting the University of Aizu.

References

1. Erl, T.: SOA Design Patterns. Prentice Hall (2010)
2. Kuniavsky, M.: Smart Things: Ubiquitous Computing User Experience Design. Elsevier (2009)
3. Folding@home Distributed Computing, http://folding.stanford.edu/
4. Beberg, A., Ensign, D., Jayachandran, G., Khaliq, S.: Folding@home: Lessons from eight years of volunteer distributed computing. In: Proc. of IEEE International Symposium on Parallel & Distributed Processing (IPDPS), Rome, Italy, pp. 1–8 (2009)
5. Cai, X., Langtangen, H.P.: Making Hybrid Tsunami Simulators in a Parallel Software Framework. In: Kågström, B., Elmroth, E., Dongarra, J., Waśniewski, J. (eds.) PARA 2006. LNCS, vol. 4699, pp. 686–693. Springer, Heidelberg (2007)
6. Ganeshamoorthy, K., Ranasinghe, D., Silva, K., Wait, R.: Performance of Shallow Water Equations Model on the Computational Grid with Overlay Memory Architectures. In: Proc. of the Second International Conference on Industrial and Information Systems (ICIIS 2007), pp. 415–420. IEEE Press, Sri Lanka (2007)
7. George, D.: TsunamiClaw User's Guide, http://faculty.washington.edu/rjl/pubs/icm06/TsunamiClawDoc.pdf/pubs/icm06/TsunamiClawDoc.pdf
8. Shuto, N., Imamura, F., Yalciner, A.C., Ozyurt, G.: TUNAMI N2; Tsunami modelling manual, http://tunamin2.ce.metu.edu.tr/

9. Titov, V.: Numerical Modeling of Tsunami Propagation by using Variable Grid. In: Proc. of the IUGG/IOC International Tsunami Symposium, Computing Center Siberian Division USSR Academy of Sciences, Novosibirsk, USSR, pp. 46–51 (1989)

10. Titov, V., Gonzalez, F.: Implementation and Testing of the Method of Splitting Tsunami (MOST). Technical Memorandum ERL PMEL-112, National Oceanic and Atmospheric Administration, Washington DC (1997)

11. Borrero, J.C., Sieh, K., Chlieh, M., Synolakis, C.E.: Tsunami Inundation Modeling for Western Sumatra. Proc. of the National Academy of Sciences of the USA 103(52) (2006), http://www.pnas.org/content/103/52/19673.full

12. Vazhenin, A., Hayashi, K., Romanenko, A.: Service-oriented tsunami wave propagation modeling tools. In: Proc. of the Joint International Conference on Human-Centered Computer Environments (HCCE 2012), Aizu-Wakamatsu, Japan, pp. 131–136. ACM Publisher (2012)

13. Vazhenin, A., Lavrentiev, M., Romanenko, A., Marchuk, A.: Acceleration of Tsunami Wave Propagation Modeling based on Re-engineering of Computational Components. International Journal of Computer Science and Network Security 13(3), 24–31 (2013)

14. iisee.kenken.go.jp/staff/fujii/OffTohokuPacific2011/tsunami.html

15. Wasielewska, K., Drozdowicz, M., Ganzha, M., Paprzycki, M., Attaui, N., Petcu, D., Badica, C., Olejnik, R., Lirkov, I.: Negotiations in an Agent-based Grid Resource Brokering Systems. In: Trends in Parallel, Distributed, Grid and Cloud Computing for Engineering. Saxe-Coburg Publications, Stirlingshire (2011)

16. Kuranowski, W., Ganzha, M., Gawinecki, M., Paprzycki, M., Lirkov, I., Margenov, S.: Forming and managing agent teams acting as resource brokers in the grid–preliminary considerations. International Journal of Computational Intelligence Research 4(1), 9–16 (2008)

17. Drozdowicz, M., Ganzha, M., Wasielewska, K., Paprzycki, M., Szmeja, P.: Using ontologies to manage resources in grid computing: Practical aspects. In: Ossowski, S. (ed.) Agreement Technologies. Law, Governance and Technology Series, vol. 8, pp. 149–168. Springer (2013)

18. Drozdowicz, M., Ganzha, M., Paprzycki, M., Szmeja, P., Wasielewska, K.: Onto-Play – a flexible user-interface for ontology-based systems, http://ceur-ws.org/Vol-918/111110086.pdf

19. Fridman, A.M., Alperovich, L.S., Shemer, L., Pustilnik, L.A., Shtivelman, D., Marchuk, A.G., Liberzon, D.: Tsunami wave suppression using submarine barriers. Physics–Uspekhi 53(8), 809–816 (2010)

20. Vazhenin, A., Watanobe, Y., Hayashi, K., Drozdowicz, M., Ganzha, M., Paprzycki, M., Wasielewska, K., Gepner, P.: Agent-based resource management in Tsunami modeling. In: 2013 Federated Conference on Computer Science and Information Systems (FedCSIS), pp. 1047–1052 (2013)

A Dataflow Platform for In-silico Experiments Based on Linked Data

Paolo Bottoni and Miguel Ceriani

Sapienza, University of Rome, Italy
{bottoni,ceriani}@di.uniroma1.it

Abstract. A big part of the work carried out by scientists nowadays involves data manipulation: access and integration of different local and online datasets, filtering for relevant information, aggregating and visualising the data to look for or defend hypotheses. The Linked Data Initiative is pushing dataset maintainers to publish data in a highly reusable way through a set of open standards, such as RDF and SPARQL. The adoption of these technologies in the scientific community is still marginal, partly for the limits of the available tools for consuming and manipulating data. We present a concrete pipeline language and a working prototype (including a visual editor and a pipeline engine) by which users can build and share applications consuming and visualising linked data.

1 Introduction

Scientists spend a significant amount of time and effort in using and configuring software tools, as part of experiment workflow, or for informal information gathering. This represents a significant burden, since scientists are not necessarily proficient programmers. Moreover most of the work is rarely used outside a single project, almost never outside a single research group.

To reduce the burden for scientists, several software tools and libraries have been developed, which simplify common tasks in specific scientific communities. Typically, a scientist uses a collection of tools, each offering one specific feature or a few of them. Such tools are used for data harvesting, data manipulation and data visualization. The work of the scientist is mainly *just* to adapt and glue together these tools, coordinating them and converting data formats as needed. It would then be beneficial to make a system available in which they can conduct such integration activities in a seamless way.

To this end, the first problem that must be faced is to assure that the envisaged system meets all the requirements of the different tools. A solution to this problem is the so called Service Oriented Architecture (SOA): instead of being delivered as software programs or libraries, specific features (called Web Services) are offered by Web servers that execute them on demand and with their resources. Web services are especially used for data manipulation, but also for data harvesting. In the latter case, the user can avoid downloading locally

A. Madaan, S. Kikuchi, and S. Bhalla (Eds.): DNIS 2014, LNCS 8381, pp. 112–131, 2014.

the remote dataset, relying on a set of Web services that execute some predefined (parametric) queries on the dataset returning a subset of it. This approach has the limit of being tied with the specific set of available queries. Moreover, both for data manipulation and for harvesting, an issue with SOA is that the workflow lifecycle is tied to the maintenance of the used services. For data visualization the trend is to use visualization tools based on Web client standards, thus accessible to any platform through a Web browser.

Another problem arises when one or some of the tools/services do *almost exactly* what is needed, i.e. the feature must be adapted to the needed requirements. If there is no access to the source code the only options are to rewrite the feature from scratch or to ask the original writer/maintainer for a customization. Even if scientists have access to the source code they must be prepared to deal with a variety of languages, libraries, and coding styles.

Finally, one deals with the problem of data format conversion, aggravated by the lack of formal definitions for some formats and the fact that custom-defined formats are usually not easily extensible when the requirements change. A basic requirement for data formats is that data should be inspectable without a need for specific tools: this can be accomplished using text-based standards, optionally compressed with standard algorithms for delivery. The need to not reinvent the wheel each time led to *meta-languages* like XML and RDF, carrying implicit structural information, and the possibility to express formally specific constraints (using XML Schema for XML and RDFS or OWL for RDF).

When the workflow is finally composed, it would be beneficial for users to have a way to store it for future reuse, as well as to be able to share it with their research group or with their broader research community: this allows the reproduction of the published experimental results and reduces the burden on colleagues aiming at improving the results or experimenting on similar ideas. A number of different systems have been developed for creating scientific workflows, usually with a pipeline approach, relying mainly on remote services for data harvesting and data manipulation.

We propose here a different approach, and a supporting platform, for the construction of scientific workflows. In this approach applications are built via cascaded definition of *views* of local or remote data, without tying execution to specific remote machines (as with Web services). The platform allows the construction of complete applications including visualization features and/or local data creation/manipulation. Throughout the system, RDF and other Semantic Web standards are used extensively for interoperability and extensibility.

Table 1 shows the RDF prefix bindings used in the paper.

In the rest of the paper, Sect. 2 introduces the basic platform requirements. After discussing technology background in Sect. 3 and related work in Sect. 4, Sect. 5 describes the proposed platform. Sect. 6 describe the language, whose semantics are specified in 7. Sect. 8 presents the user interface of the editor, while Sect. 9 presents the software implementation and Sect. 10 outlines an example application. Finally, Sect. 11 discusses conclusions and future work.

Table 1. RDF prefix bindings used in the paper

Prefix	IRI
rdf:	http://www.w3.org/1999/02/22-rdf-syntax-ns#
rdfs:	http://www.w3.org/2000/01/rdf-schema#
xsd:	http://www.w3.org/2001/XMLSchema#
df:	http://www.swows.org/dataflow#
xml:	http://www.swows.org/2013/07/xml-dom#
evt:	http://www.swows.org/2013/07/xml-dom-events#
nn:	http://www.swows.org/xml/no-namespace#
svg:	http://www.w3.org/2000/svg#
xlink:	http://www.w3.org/1999/xlink#
swi:	http://www.swows.org/instance#
geo:	http://www.fao.org/countryprofiles/geoinfo/geopolitical/resource/
map:	http://localhost/DemoWorld/BlankMapWithRadioBox.svg#
:	http://www.swows.org/samples/WorldInfo#

2   Proposed Platform Requirements

By analysing literature and common practices in scientific workflows, especially in the biomedical and the health fields, we have elicited a set of requirements, reflecting both common practices and desired advancements. As a consequence, we demand that the platform:

- be based on a *dataflow language* in which data trasformations are represented as pipelines;
- use pipelines through cascading *declarative views* on the input or other views;
- represent data as *RDF*;
- be able to connect to existing *RDF sources*;
- exhibit *interactivity* through Web interface input/output;
- represent pipelines *as RDF* to share and re-use;
- support interoperability with *XML*;
- use *existing standards* whenever possible.

3   Technologic Background

Most of the technologies adopted in our project are based on the Extensible Markup Language (XML) [1], which is used both for the definition of new languages, i.e. acting as a meta-language, and as a language for the expression of structured information.

3.1   RDF

The relational model is widely used to represent virtually any kind of structured information. The Resource Description Framework (RDF) [2] generalises it to the universe of structured data in the World Wide Web, better known as the Semantic Web [3]. In the RDF data model, knowledge is represented via statements about resources, where a resource is an abstraction of any piece of information about some domain. A RDF statement (also called a triple) is composed of *subject* (a resource), *predicate* (specified by a resource as well) and *object* (a resource or a literal, i.e. a value from a basic type). A RDF graph is therefore a set of triples. Resources in a graph can be identified by a URI if they have meaning outside of that graph, or by a local identifier otherwise (in which case they are called blank nodes). The resources used to specify predicates are called properties. A resource may have one or more types, specified by the predefined property rdf:type. A RDF dataset[1] is a set of graphs, each associated with a different name (a URI), plus a default graph without a name. We use RDF through the framework to represent any kind of information and its transformations.

Blank Nodes. Blank Nodes is a feature present in RDF allowing the representation of structures in which not every node needs to be referenceable on its own (and thus have an URI). This enables simple representations of nodes that are needed in the structure of an RDF Graph but have no specific meaning outside of it. Such nodes, however, are often seen as problematic due to distinct conflicting interpretations in their actual use and to the added complexity they bring to different operations on RDF graphs [6].

3.2   SPARQL

We extensively use SPARQL [4], the standard query language for RDF datasets. SPARQL has a relational algebra semantics, analogous to those of traditional relational languages, such as SQL. The SPARQL *construct*, one of the possible SPARQL query forms, takes as input a RDF dataset and produces a RDF graph. While the SPARQL Query Language is "read-only", the SPARQL Update standard [7] defines a way to perform updates on a RDF Graph Store, the "modifiable" version of a RDF Dataset. A SPARQL Update *request* is composed of a number of SPARQL Update *operations*.

The current version of the standard is SPARQL 1.1, but we mention also the previous version, SPARQL 1.0 [8], because much of the existing work refers to that version. SPARQL 1.1 algebra offers a much expanded set of operators, effectively allowing the expression of queries that were not expressible before.

3.3   Rich Web Clients

The ubiquity of Web browsers and Web document formats across a range of platforms and devices drives developers to choose to build applications on the Web and its standards. From the point of view of the requirements for browser, there has been a dramatic change from the first days of the Web. Now a

---

[1] Defined in SPARQL [4], but to be part of the core RDF standard from RDF 1.1 [5].

browser is an interface to an ever growing set of client capabilities. All modern browsers natively support the Scalable Vector Graphics (SVG) standard [9], an XML-based language representing mixed vector and raster content. Together with the long established Document Object Model (DOM) Events [10,11] and JavaScript support, it allows the realisation of complete interactive visualisation applications. Indeed, JavaScript libraries for interactive data visualization are proliferating, from standard visualisations [12,13,14] to specialized visualisations for specific domains [15], leveraging especially on SVG technology.

4 Related Work

We report on a number of topics central to our project, in particular with reference to the principal features of the languages involved in it.

4.1 View Definition Languages

There have been different proposals of languages to define SPARQL views, in a way analogous to SQL views. A SPARQL view is a graph intensionally defined through a SPARQL Construct query; the input dataset of the query can be composed by both "real" graphs (extensionally defined) and views.

RVL [16] is an early effort, using an imperative language for defining views based on an independently defined query language (RQL [17]). Views are associated with "namespaces" (identified through unique URIs), representing virtual RDF/S datasets. Namespaces can be both extensionally and intensionally defined, as RQL queries of other namespaces. The query language offers some relations and operators to deal with RDFS concepts (rdf:type, rdfs:subClassOf, rdfs:subPropertyOf, rdfs:domain, rdfs:range). vSPARQL [18] is an extension of SPARQL 1.0 grammar allowing named views defined with Construct queries and reusable in other queries. Schenk and Staab, working on *Networked Graphs* [19], propose an RDF-based syntax to define views, which are graphs defined in terms of SPARQL 1.0 Construct queries on explicitly defined graphs and other views.

Cyclic Views. The structure formed by the defined views form a graph that could in general be cyclic. While in RVL cycles are avoided, vSPARQL and Networked Graphs allow cycles as a way to recursively define views, increasing the expressive power of the query language[2].

The possibility of recursion comes at a cost: it is not possible to guarantee termination. There are also different possible semantics that can be defined, related to fixpoint semantics in logic. As in SPARQL (even in version 1.0), it is possible to use negation in queries [19]; queries can be non-monotonic.

---

[2] A simple example is the transitive closure of a property, that could not be expressed in SPARQL 1.0 without recursion. In SPARQL 1.1 simple transitive closure can be directly achieved, but this does not hold for slightly more complex examples.

The approach of vSPARQL is to force recursive queries to be monotonic by allowing only the so-called *stratified negation* [20], used also to define recursion in SQL[21]. Each recursive query is statically checked before execution and an error is raised if non-stratified negation is found. The least fixpoint is then computed by iterative application of the query (starting from the empty graph).

In Networked Graphs there is no such control. Instead, all kinds of (SPARQL 1.0 based) recursive queries are permitted, but interpreted by the so-called *well-founded semantics* [22]. These are three-valued semantics (*true, false, unknown*) and in order to use them the SPARQL semantics are slightly modified to allow the indetermination of part of the recursion defined graph. There are different non-stratified negation examples in which this approach converges, in some cases to a complete model (there are no *unknown* conclusions in the end), in other cases to a partial model [23]. In Networked Graphs the solution graph is chosen to be formed by the triples proven to be true (the distinction between false and unknown triples is not further propagated after recursion, as RDF lacks an easy way to represent three-valued semantics).

Although powerful enough to define read-only applications (possibly together with visualization tools described below), network of views do not easily model interactive applications. In particular, they face the problem of how to represent events and time-dependent information, including the application state.

4.2 Pipeline Languages and Linked Data Visualization Tools

Two pipeline languages have been proposed to define RDF transformations, namely DERI Pipes [24] and SPARQLMotion [25]. They offer a set of basic operators on RDF graphs to build the pipelines, which are then typically executed in the context of a batch or Web application (within a Web site or a Web service), in response to GET or POST requests. They are also both endowed with a graphical environment to create the pipelines using the available operators (free in the case of DERI Pipes, in a commercial software[3] for SPARQLMotion).

In DERI Pipes, pipelines are defined in XML, with SPARQL queries in textual form. The set of operators includes one that loads RDF graphs from files or as result of a query to a SPARQL End Point (FETCH), one that transforms RDF graphs using SPARQL Construct queries (CONSTRUCT) and one that merges a set of RDF graphs into one (MIX). The operators have no side-effects and the pipelines are stateless.

In SPARQLMotion, both pipelines and queries are represented in RDF (for queries using the SPIN-SPARQL [26] syntax). SPARQLMotion has operators similar to DERI Pipes together with a number of operators converting to and from a number of data formats (XML among them) and RDF. The main difference with DERI Pipes is the presence of operators with side-effects, e.g. to save to file, update the *active* RDF Graph, or send emails.

---

[3] http://www.topquadrant.com/products/TB_Composer.html

Visualbox [27] and Callimachus [28] have been proposed explicitly for linked data visualisation. In their two-step model/view approach, SPARQL queries select data and a template language generates the (XML-based) visualisation.

SPARQL Web Pages (SWP) is a RDF-based framework (to be used with SPARQL Motion or on its own) to describe and render HTML+SVG visualisations of linked data. HTML and SVG are mapped to two corresponding vocabularies and together with the UISPIN Core Vocabulary allow the association of a RDF resource with the description of its visualisation. The description may be also statically associated with a class of resources, with each specific resource mapping defined through a SPARQL query (in SPIN-SPARQL syntax).

In all these proposals the execution model corresponds to the management of a single HTTP request, as with typical application server technologies like Java Servlet or PHP. Persistence and logical relationships between requests and client state must be managed explicitly (e.g. saving/loading data related to a session and encoding parameters in requests)[4].

4.3 RDF-Based Transformation Languages

We already presented Networked Graphs and SPARQLMotion, two languages based on RDF. Other proposals for RDF-based declarative languages operating on RDF include the Ripple [29] language, functional and stack-oriented, which takes a resource centric approach (in constrast with the SPARQL relational approach) and a language to apply Algebraic Graph Transformations to RDF Graphs [30]. These are interesting alternative approaches, but from the point of view of the developer they are impervious because of the need to learn from scratch a non-standard language. We think that in order to push the boundaries of the existing standards (especially SPARQL) in a pragmatical way, we need to integrate and extend them while preserving interoperability.

5   Proposed Platform

The platform we propose allows users to define linked data applications through a pipeline language based on RDF (see Sect.6). Users will be able to create pipelines in a modular way and using a visual representation, through a Web-based *editor* (see Sect.8), in turn interacting with a *pipeline repository*. Another software, called the *dataflow engine*, will execute the pipeline (after download-ing the corresponding RDF Graph from the repository) on a possibly separate server, with a Web-based interface as well. Figure 1 shows a simplified lifecycle of the pipeline, from editing and saving it in a directly controlled repository, to eventually sharing it for use "as it is" or reuse in other pipelines.

---

[4] Both Callimachus and SWP offer some aid for building interactive applications via special functions and syntaxes, but the execution model remains request oriented.

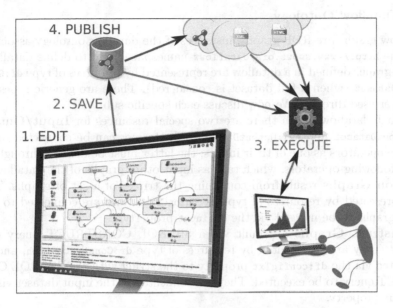

Fig. 1. A schematic view of the platform workflow

6   Language Description

We propose a declarative language for RDF-based applications, called Dataflow
Pipeline Language (DfPL).

As DfPL is aimed at creating applications and modules that can be shared
and reused, any module should be as independent and self contained as possible.
To achieve this goal, DfPL is built using operators free from side-effects.

While languages considered in Sect. 4 are either stateless or have operators
with side-effects that modify the system state, e.g. updating some RDF Store,
DfPL offers a special operator (*Updatable Graph*) to define a portion of the
application state. This operator defines a virtual graph in terms of a set of
triggers: the graph is updated when other certain graphs change. Hence we adopt
a model of state which is not centralized, but distributed and bound directly to
the related functional module.

In DfPL the basic functional module is the *Dataflow*. A Dataflow defines how
to build an output RDF dataset from an input one, through a configuration RDF
graph, called the *dataflow graph*. The content of the output dataset is defined as
the result of the cascaded application of graph operators to the input dataset.

A pipeline can be designed just for reuse by other pipelines. If a pipeline has to
be executed (i.e. it is a *top-level* pipeline), its default output graph must comply
with an *XML DOM Ontology*(see 6.2) describing the XML DOM in RDF. It will
represent a HTML or SVG document, to be rendered by the user interface. Its
default input graph will receive the DOM Events generated in the user interface,
described with a *DOM Events Ontology*(see 6.3).

6.1   Dataflow Ontology

Dataflow graphs are RDF graphs, instances of the dataflow ontology associated with the http://www.swows.org/dataflow# namespace, used to define dataflows.

The graphs defined in a dataflow are represented by resources of type df:Graph or df:Dataset (when a full dataset is considered). These are generic types and are never used directly. We now discuss each specific subtype.

In each dataflow graph there are two special resources for **Input/Output**: swi:InputDataset and swi:OutputDataset. The former can be connected to one or more operators as one of their inputs; the latter must be defined through one of the following operators, which take as input other graphs of the dataflow.

Union Graphs result from combining the triples of a set of graphs. They are represented by resources of type df:UnionGraph, and are connected to each of the graphs to be merged via the df:input property.

Construct Graphs are built via a SPARQL CONSTRUCT query on a dataset. They are represented by resources of type df:ConstructGraph, and are connected via the df:configTxt property to the string with the SPARQL CON-STRUCT query to be executed. They are connected to the input dataset via the df:input property.

Inside a dataflow graph, other (sub)dataflows, called **Inner Dataflows**, can be applied. They are represented by resources of type df:DataflowGraph, and are connected to their dataflow graph via the df:config property. Like the construct graphs, they are connected to the input dataset via the df:input property.

Updatable Graphs are the only ones having memory. They are represented by resources of type df:UpdatableGraph, and are connected via the df:configTxt property to a string with a SPARQL Update request, possibly composed of more than one operation. They are connected to the input dataset via the df:input property. Initially the content corresponds to the empty graph. Each time one of the input graphs changes, the Update request is executed on the current content.

Datasets are built from graphs with df:InlineDataset, while df:SelectGraph extracts a graph from a dataset.

To represent XML and any dataset containing a special node that must be identified as such, the URI swi:GraphRoot is used.

6.2   XML DOM Elements

As many of the intended applications of DfPL rely on XML formats, RDF graphs are used also to represent XML documents. Hence, an ontology[5] has been defined to describe XML documents in RDF, based on the concepts in the XML Document Object Model (DOM) [31].

The ontology presents classes such as xml:Element, xml:Attr, xml:Text to represent the basic XML node types. By subclassing them or using them directly it is possible to represent the XML nodes. The engine provides also some methods to avoid explicitly subclassing the known elements, so that they can be used directly (see 10.7).

---

[5] http://www.swows.org/2013/07/xml-dom

The hierarchical relation between DOM nodes is represented using the property `xml:hasChild`. In case the relative order of the children in the DOM tree is important, there are two ways to express it: 1) using the `xml:orderKey` property on each child we want to order, or 2) using the `xml:childrenOrderedBy` property on the parent. This latter property has another property as range, this one being the property that will be used to order its children. The order occurs in both cases following the semantics of the operator < in SPARQL and can be reversed to descending order using the `xml:childrenOrderType` property with the value `xml:Descending` on the parent. This is convenient from the point of view of the SPARQL developers as they can flexibly define orders, instead of being forced to indicate the exact index of a child or to create a linked list of children. Moreover, in this model a single RDF node can be used as child of different parents, and thus converted into different XML DOM nodes (this can be useful if the same DOM subtree is needed in different places of a document) that can be even ordered in different ways (depending on the properties of each parent). If an element uses an `id` attribute, this corresponds in the RDF model to a resource with the URI corresponding to the fragment identifier syntax in XML. This allows easy reference from SPARQL queries to specific elements inside a XML document.

6.3   Event Streams

In order to bridge SWOWS to existing event management infrastructures, an ontology[6] has been defined representing DOM Events [10], from generic to specific ones, e.g. mouse events, mimicking the interfaces of the W3C DOM standard.

In the output, each node of type `xml:Element` can have one or more values for the property `xml:listenedEventType`, describing the event types that the application needs to listen to.

7   Language Semantics

We informally describe the intended semantics of the language. In the standard terminology, an *RDF source* [5] is a mutable (in time) source of RDF graphs; a snapshot of the state of a RDF source is a RDF graph. The input dataset of a dataflow is a set of named RDF sources (and a default one); or we can say that this dataflow is "applied" to a set of RDF sources. The result (represented by the output dataset) is another set of named RDF sources.

Now we want to consider just the stateless (combinatorial) part of a dataflow. Thus we consider each of the Updatable graphs as part of both the input and the output set of RDF sources, where in the output it represents the new state after the application of an update request. The update can be considered a stateless operation that generates the new state of the RDF graph from its previous state and the other inputs. The stateless network can calculate the state of the output

---

[6] http://www.swows.org/2013/07/xml-dom-events

RDF sources as a function of the state of the input RDF sources, after possibly converging in the case of cycles (see 7.1 for some details).

We consider the variation in time of the input RDF sources as discrete, this is compatible with user interface event models. We call the start of the system and any variation of the RDF sources a *change event*. For any input change event there can be an output change event (or not if the output state is unchanged). This output change event may include some of the Updatable graphs, in that case inducing a new input change and in turn leading to recalculate a new output state. This iteration happens until the output state of the Updatable graphs is unchanged and before considering other "external" change events. To provide a model for user interface events we consider also istantaneous events, i.e. a content that has meaning only in that snapshot. In that case the event graph is available only for the first step.

7.1 Cycles

Cycles in the dataflow are permitted, so that it is possible to define views recursively, as in vSPARQL and Networked Graphs. For the semantics of cyclic definitions we chose an approach halfway between vSPARQL and Networked Graphs. We allow non-stratified negation, thus allowing recursive queries that could not be expressed in vSPARQL. Differently from Networked Graphs we do not change the SPARQL semantics, sticking to the *negation as failure* interpretation. Thus there are cases in which the Networked Graphs would converge to a *partial* well-founded semantics model [23], while our model does not converge, since it does not allow indetermination in the final represented graph. The algorithm starts from the empty set and iteratively applies the chain of operators composing the cycle until the output is a graph identical to the input. This approach finds the least (if existing) fixpoint in the case of stratified negation [20] and finds (if it stops) a (complete) stable model in the general case.

8   Editor User Interface

The editor is contained in a Web page (see Fig. 2) providing tools to create and modify dataflows saved as RDF graphs on a Graph Store [32] and composed of:

- a *component panel*(left), where components represent language operators;
- the *data sources tab* (left), allowing reference to RDF data sources already known to the system or the creation of new ones;
- the *pipelines tab* (left), allowing developers to use, view or modify other pipelines created by the user or any pipeline available online (in which case it would not be modifiable);
- the *editor area* (center), where the pipeline is built by dragging, linking and configuring the desired components;
- the *command panel* (upper right), contains some buttons for operations related to the whole pipeline, e.g. saving it or executing it;
- the *helper area* (bottom-left corner), for contextual help on components;
- the *source code area* (bottom), showing the RDF graph for the dataflow.

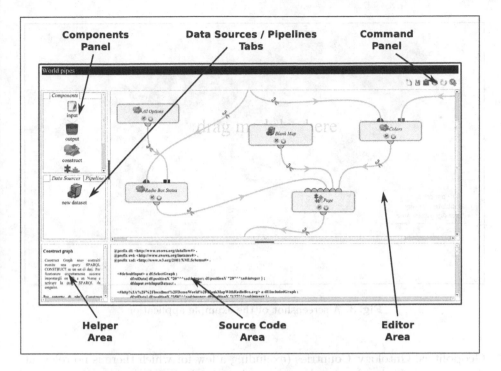

Fig. 2. Main interface elements of the Web-based editor

9 Implementation

The main blocks of the application are the *editor*, the *pipeline repository* and the *dataflow engine*. The editor is a rich Web application with its client side logic coded in HTML+CSS+JavaScript. The pipeline repository is an instance of Graph Store that must be located in the same host as the editor. The dataflow engine is a Java based (using Apache Jena [33]) Web application maintaining the state of each running pipeline instance; when a new instance is launched (e.g., from the editor) the engine initialises the pipeline and returns its output to the client, along with a piece of JavaScript logic to report handled events back to the server; each time an event is fired on the client, the dataflow engine is notified and answers with the changes to be executed on the client content. On the client side, any *modern* browser supporting JavaScript can use both the editor and the generated application. The software is free and available on line[7].

10 A Toy Example: Visualisation on a Map

We will demonstrate how to build a data visualization through the creation of a pipeline[8]. Our aim is to show on a world map some data from the FAO

---

[7] http://www.swows.org/

[8] A clip of this example is available at http://youtu.be/_PrOyYU2E0g

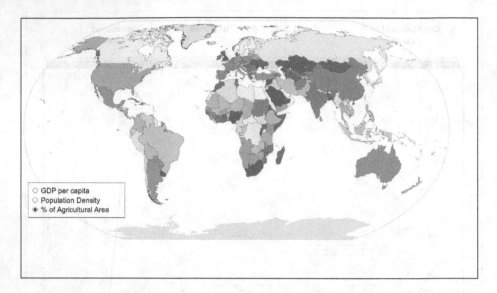

Fig. 3. A screenshot of the example application

Geopolitical Ontology. Countries (excluding a few for which there is no relevant data in the ontology) are coloured on the map in a colour depending on an index value. The index used can be chosen from three different options: *GDP per capita, population density* and *proportion of agricultural area*. Figure 3 is a screenshot of the example application.

10.1   The Blank Map

One of the platform aims is to be interoperable with other Web content, in order to favour reuse of resources. Thus we are using one of the Wikipedia blank world maps. They are freely available SVG files to be used for creating static or interactive visualizations upon them. We can refer remotely to the SVG file and use it as it is in a pipeline, dynamically changing it to create the desired visualisation. For this example, however, we decided to download it and change it, adding the static parts that are needed in our appplication. The modified SVG contains a box with radio buttons and labels for the available options. The file contains also the definition of the black circle to be shown inside the selected radio button; this definition is not used in the blank SVG but it will help to keep the dynamic part lighter. The modified file must be uploaded to some local or online server to be available[9]; in this example we uploaded it to the local web server. The file will be finally referenced in the pipeline as a Data Source and "tunneled" in it as RDF (e.g., to allow its modification by SPARQL).

---

[9] The upload is executed outside of the user interface but in the future we plan that upload of files to the same server will be integrated with the pipeline interface.

10.2 FAO Ontology

The FAO Geopolitical Ontology has been developed to standardise information about countries and/or regions and is freely available[10]. Even if its main aim is to provide a reference for different codes and names of countries and regions, there are a few general statistics related to the countries. Our aim is to extract the desired geopolitical indexes from the data. In our system there are two ways to access linked data: accessing as Data Source a file containing RDF (in one of different syntaxes), accessing a SPARQL Endpoint through the SPARQL Federation syntax. Since a source file must be completely downloaded if data is to be accessed from it, the preferred way is to access a SPARQL Endpoint. As in the case of the FAO Geopolitical Ontology there is no SPARQL Endpoint, it is only available for download as a single RDF. For convenience we uploaded it as a RDF Graph on the local Graph Store, so that it can be accessed with the local SPARQL Endpoint. The following is the query used in the pipeline to extract the indexes from the original data. Here and in the following queries we omit the prefixes declaration for brevity (see Table 1 for the bindings).

```
 1  CONSTRUCT {
 2    ?country
 3      a geo:self_governing ;
 4      geo:codeISO2 ?codeISO2upper ;
 5      :gdpPerCapita ?gdpPerCapita ;
 6      :density ?density ;
 7      :agriAreaRate ?agriAreaRate .
 8  }
 9  WHERE {
10    SERVICE <http://localhost:3030/ds/sparql> {
11      GRAPH geodata: {
12        ?country
13          a geo:self_governing ;
14          geo:codeISO2 ?codeISO2upper ;
15          geo:GDPTotalInCurrentPrices ?gdp ;
16          geo:populationTotal ?pop ;
17          geo:landAreaTotal ?landArea ;
18          geo:agriculturalAreaTotal ?agriArea .
19      }
20    } .
21    BIND( ?gdp / ?pop AS ?gdpPerCapita ) .
22    BIND( ?pop / ?landArea AS ?density ) .
23    BIND( ?agriArea / ?landArea AS ?agriAreaRate ) .
24  }
```

The empty prefix (:) is used as a convenience local namespace for all resources that have no meaning outside of the pipeline. The query simply derives the value of the three indexes for each country. It is to be noted that the properties :gdpPerCapita, :density and :agriAreRate will be used in other queries as subject or objects of triples, in order to specify the chosen index.

10.3 Pipeline

The pipeline, shown in Fig. 4, is composed by the following components:

- the Data Source *Blank Map* that loads the SVG Blank Map described in 10.1;

---

[10] http://www.fao.org/countryprofiles/geoinfo/geopolitical/resource/

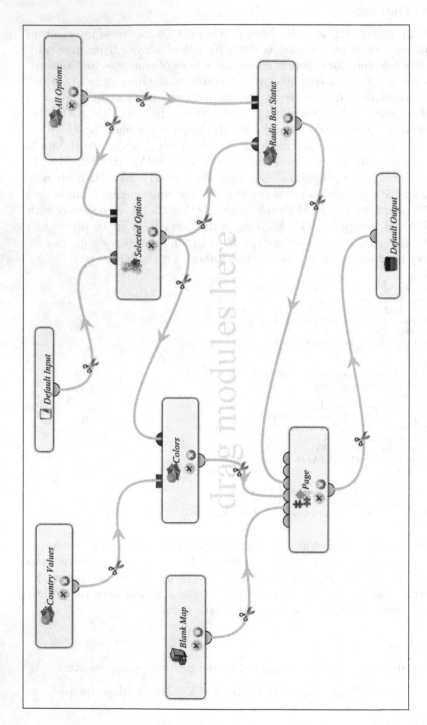

Fig. 4. The pipeline of the example application

- the Construct graph *Country Values*, whose query is described in 10.2, that extracts the indexes from the FAO Geopolitical Ontology;
- the Construct graph *All Options* (see 10.4), that generates the set of (three) possible options;
- the Updatable graph *Selected Option* (see 10.5), that sets the initial selected option and change it when an another radio button is pressed (events coming from *Default Input*);
- the Construct graph *Colors* (see 10.6), that generates the styles needed to colour each country, taking from Country Values the data related to the Selected Option;
- the Construct graph *Radio Box Status* (see 10.7), that generates the dynamic part of the radio buttons box, depending on the Selected Option;
- the Union graph *Page*, that merges the static part of the visualization (Blank Map) with the dynamic ones (Colors and Radio Box Status), creating the *Default Output* graph that will be transformed in an SVG view (in this case, in general it would be HTML+SVG).

10.4   Available Options

We need a sort of configuration of the available options to choose from, as well as of the association with the related visual elements in the SVG map (the radio buttons). While a configuration RDF file could have been used, also static content can be created with a SPARQL query. Thus we use a Construct component with the following query.

```
 1  CONSTRUCT{
 2    ?optionID  :optionElement ?optionElement .
 3  }
 4  WHERE{
 5    VALUES (?optionID ?optionElement) {
 6      (:gdpPerCapita map:option_gdpPerCapita)
 7      (:density map:option_density)
 8      (:agriAreaRate map:option_agriAreaRate)
 9    } .
10  }
```

10.5   Selected Option

In this very simple application the state consists only of the selected option. The *Selected Option* component is thus an Updatable graph component that is updated when a new listened event is fired from user interface, i.e. when the Default Input Graph changes. The component executes the following SPARQL Update request (consisting of two operations).

```
 1  INSERT { swi:GraphRoot :selectedOption :gdpPerCapita }
 2  WHERE {
 3    FILTER NOT EXISTS{ swi:GraphRoot :selectedOption ?anyOption } .
 4  };
 5
 6  DELETE { swi:GraphRoot :selectedOption ?oldOption }
 7  INSERT { swi:GraphRoot :selectedOption ?newOption }
 8  WHERE {
```

```
 9     swi:GraphRoot :selectedOption ?oldOption .
10     GRAPH <#event> {
11       ?event
12         a evt:Event ;
13         evt:type "mousedown" ;
14         evt:currentTarget ?newOptionElement .
15     } .
16     GRAPH <#availableOptions> {
17       ?newOption :optionElement ?newOptionElement .
18     }
19     FILTER ( ?newOption != ?oldOption ).
20   };
```

The first operation (lines 1-4) is executed (when still no selection exists) to set up the initial state of the graph; following this operation there will always be a selected option and thus it has effect only once. The second operation (lines 6-20), conversely, is executed each time a new option is selected: when the target of the upcoming event matches the element associated with an (unselected) option.

10.6 Colouring the Map

The *Colors* component takes the output of Country Values to color the map accordingly. The following SPARQL query, slightly more complex than the others, do the job.

```
 1   CONSTRUCT {
 2     map:variable_css xml:hasChild <#style_text> .
 3     <#style_text>
 4       a xml:Text ;
 5       xml:nodeValue ?style_value .
 6   }
 7   WHERE {
 8     SELECT (GROUP_CONCAT("." + ?country_class + ?country_style_value)
 9                 AS ?style_value)
10     WHERE {
11       GRAPH <#option> {
12         swi:GraphRoot :selectedOption ?option .
13       }
14       GRAPH <#countryValues> {
15         ?country
16           a geo:self_governing ;
17           geo:codeISO2 ?codeISO2upper ;
18           ?option ?value .
19         {
20           SELECT ?option (MAX(?value) AS ?maxValue)
21           WHERE {
22             ?country
23               a geo:self_governing ;
24               ?option ?value .
25           } GROUP BY ?option
26         }
27         {
28           SELECT ?option (MIN(?value) AS ?minValue)
29           WHERE {
30             ?country
31               a geo:self_governing ;
32               ?option ?value .
33           } GROUP BY ?option
34         } .
35       } .
36       BIND((?value - ?minValue)/(?maxValue - ?minValue) AS ?proportion).
37       BIND(LCASE(?codeISO2upper) AS ?country_class).
38       BIND(STR(255) AS ?red).
```

```
39   BIND(STR(xsd:integer(ROUND(xsd:decimal((1 - ?proportion) * 255))))
40        AS ?green).
41   BIND(STR(xsd:integer(ROUND(xsd:decimal((1 - ?proportion) * 255))))
42        AS ?blue).
43   BIND(
44      CONCAT(
45        "{fill: rgb(", ?red, ",", ?green, ",", ?blue,") !important}")
46           AS ?country_style_value) .
47   }
48 }
```

The option from Selected Option (lines 11-13) is used to chose the property to read from each country in Country Values. To normalize values of the selected property in the range 0-1, minimum and maximum values are calculated in two subqueries (lines 19-26 and 27-34) and used to calculate the normalized value (line 36). From that value the RGB components of the color (shades of red) are calculated (lines 38-42) and then the style assignment for each country is built (lines 43-46). All these style assignments are finally aggregated in a string (lines 8-9), that is attached as text child of a style element of the SVG document (lines 2-5).

10.7 Radio Box Status

The *Radio Box Status* component defines the dynamic aspects of the radio buttons box. It is a Construct graph with the following query.

```
1  CONSTRUCT {
2     ?selectedOptionElem xml:hasChild <#selectedOption_use> .
3     <#selectedOption_use>
4        a svg:use ;
5        xml:orderKey 2 ;
6        xlink:href "#radioPoint" .
7     ?unselectedOptionElem xml:listenedEventType "mousedown" .
8  }
9  WHERE {
10    GRAPH <#option> {
11      swi:GraphRoot :selectedOption ?selectedOption .
12    }
13    GRAPH <#availableOptions> {
14      ?selectedOption :optionElement ?selectedOptionElem .
15      ?unselectedOption :optionElement ?unselectedOptionElem .
16    }
17    FILTER( ?selectedOption != ?unselectedOption ) .
18 }
```

It places the black circle (referenced in the SVG document as `#radioPoint`) inside the element corresponding to the selected option. For all the other options the pipeline listens to `mousedown` events.

11   Conclusions and Future Work

We have presented a platform to build pipelines which specify transformations of RDF graphs in order to build data manipulation and visualisation applications. The RDF pipeline language is based on existing standards (such as SPARQL) and is unique in having been designed for interactive applications and thus able to react to graph modification events.

We are proposing this system as a proof-of-concept and as a testbed for experimentation in RDF programming with a dataflow approach. We want to leverage this experimentation to build higher level interfaces, designed also for usage by non-expert users, as a way to flexibly interact with linked data.

References

1. Bray, T., Paoli, J., Sperberg-McQueen, C.M., Maler, E., Yergeau, F., Cowan, J.: Extensible Markup Language (XML) 1.1 (Second Edition). W3C Recommendation 16 August 2006, edited in place 29 September 2006
2. Klyne, G., Carroll, J.J., McBride, B.: Resource Description Framework (RDF): Concepts and Abstract Syntax. W3C Recommendation (February 10, 2004)
3. Berners-Lee, T., Hendler, J., Lassila, O.: The semantic web. Scientific American 284(5), 34–43 (2001)
4. Harris, S., et al.: SPARQL 1.1 Query Language. W3C Recommendation (March 21, 2013)
5. Cyganiak, R., Wood, D., Lanthaler, M.: RDF 1.1 Concepts and Abstract Syntax. W3C Candidate Recommendation (November 5, 2013)
6. Mallea, A., Arenas, M., Hogan, A., Polleres, A.: On Blank Nodes. In: Aroyo, L., Welty, C., Alani, H., Taylor, J., Bernstein, A., Kagal, L., Noy, N., Blomqvist, E. (eds.) ISWC 2011, Part I. LNCS, vol. 7031, pp. 421–437. Springer, Heidelberg (2011)
7. Schenk, S., Gearon, P., et al.: SPARQL 1.1 Update. W3C Recommendation (March 21, 2013)
8. Prud'hommeaux, E., Seaborne, A.: SPARQL Query Language for RDF. W3C Recommendation (January 15, 2008)
9. Andersson, O., Armstrong, P., Axelsson, H., Berjon, R., Bzaire, B., et al.: Scalable Vector Graphics (SVG) 1.1 Specification. W3C Recommendation (January 14, 2003)
10. Pixley, T.: Document Object Model (DOM) Level 2 Events Specification. W3C Recommendation (November 13, 2000)
11. Kacmarcik, G., Leithead, T., Rossi, J., Schepers, D., Hhrmann, B., Le Hgaret, P., Pixley, T.: Document Object Model (DOM) Level 3 Events Specification. W3C Recommendation (November 13, 2000)
12. Google: Google charts (2010)
13. Belmonte, N.G.: Javascript infovis toolkit (2011)
14. Bostock, M., Ogievetsky, V., Heer, J.: D3: Data-driven documents. IEEE Trans. Visualization & Comp. Graphics, Proc. InfoVis (2011)
15. Smits, S.A., Ouverney, C.C.: jsPhyloSVG: A Javascript Library for Visualizing Interactive and Vector-Based Phylogenetic Trees on the Web. PloS one 5(8), e12267 (2010)
16. Magkanaraki, A., Tannen, V., Christophides, V., Plexousakis, D.: Viewing the Semantic Web through RVL Lenses. In: Fensel, D., Sycara, K., Mylopoulos, J. (eds.) ISWC 2003. LNCS, vol. 2870, pp. 96–112. Springer, Heidelberg (2003)
17. Karvounarakis, G., Magkanaraki, A., Alexaki, S., Christophides, V., Plexousakis, D., Scholl, M., Tolle, K.: Rql: A functional query language for rdf. In: Gray, P.D., Kerschberg, L., King, P.H., Poulovassilis, A. (eds.) The Functional Approach to Data Management, pp. 435–465. Springer, Heidelberg (2004)

18. Shaw, M., Detwiler, L.T., Noy, N., Brinkley, J., Suciu, D.: vSPARQL: A view definition language for the semantic web. Journal of Biomedical Informatics 44(1), 102–117 (2011); Ontologies for Clinical and Translational Research
19. Schenk, S., Staab, S.: Networked graphs: a declarative mechanism for SPARQL rules, SPARQL views and RDF data integration on the web. In: Proc. WWW 2008, pp. 585–594. ACM (2008)
20. Chandra, A.K., Harel, D.: Horn clause queries and generalizations. The Journal of Logic Programming 2(1), 1–15 (1985)
21. ISO-ANSI: Database language sql-part2: Sql/foundation. Technical Report 9075-2 edition, ANSI, ISO 9075-2 edition (1999)
22. Van Gelder, A., Ross, K.A., Schlipf, J.S.: The well-founded semantics for general logic programs. Journal of the ACM (JACM) 38(3), 619–649 (1991)
23. Van Gelder, A.: The alternating fixpoint of logic programs with negation. Journal of Computer and System Sciences 47(1), 185–221 (1993)
24. Le-Phuoc, D., Polleres, A., Hauswirth, M., Tummarello, G., Morbidoni, C.: Rapid prototyping of semantic mash-ups through semantic web pipes. In: Proc. WWW 2009, pp. 581–590. ACM (2009)
25. Knublauch, H., et al.: SPARQLMotion Specifications (2010), http://sparqlmotion.org/
26. Fürber, C., Hepp, M.: Using SPARQL and SPIN for data quality management on the semantic web. In: Abramowicz, W., Tolksdorf, R. (eds.) BIS 2010. LNBIP, vol. 47, pp. 35–46. Springer, Heidelberg (2010)
27. Graves, A.: Creation of visualizations based on linked data. In: Proceedings of the 3rd International Conference on Web Intelligence, Mining and Semantics, vol. 41, ACM (2013)
28. Battle, S., Wood, D., Leigh, J., Ruth, L.: The Callimachus Project: RDFa as a Web Template Language. In: COLD (2012)
29. Shinavier, J.: Functional programs as linked data. In: 3rd Workshop on Scripting for the Semantic Web (2007)
30. Braatz, B., Brandt, C.: How to Modify on the semantic web? In: Daniel, F., Facca, F.M. (eds.) ICWE 2010. LNCS, vol. 6385, pp. 187–198. Springer, Heidelberg (2010)
31. Apparao, V., Byrne, S., Champion, M., Isaacs, S., et al.: Document Object Model (DOM) Level 1 Specification - Version 1.0. W3C Recommendation (October 1, 1998)
32. Ogbuji, C.: Sparql 1.1 Graph Store HTTP Protocol. W3C Recommendation (March 21, 2013)
33. McBride, B.: Jena: a semantic Web toolkit. IEEE Internet Computing 6(6), 55–59 (2002)

Knowledge Visualization of Reasoning
for Financial Mathematics with Statistical Theorems

Yukari Shirota[1], Takako Hashimoto[2], and Sakurako Suzuki[3]

[1] Department of Management, Faculty of Economics, Gakushuin University, Tokyo, Japan
yukari.shirota@gakushuin.ac.jp
[2] Commerce and Economics, Chiba University of Commerce, Chiba, Japan
takako@cuc.ac.jp
[3] Department of Management, Faculty of Economics, Gakushuin University, Tokyo, Japan
SuzukiSakurako@aol.com

Abstract. We have been developing formula databases for our business math lectures. Our target is business mathematics and the main part is financial mathematics. To handle financial word problems, many statistical formulas and theorems must be referenced and used. We describe visualization of the statistical formula database in the paper. To teach statistical theorems, simulation and visualization is so helpful. Therefore such visualization materials should also be stored in the formula database as a piece of our knowledge. On the other hand, the problem exists that math formula database researches may hardly consider its user interfaces. We have been researching a deductive reasoning process for solving a word math problem. Then we proposed that the reasoning process graph can be used as an effective and excellent user interface of the large-scale knowledge base for students to solve a word math problem. In the paper, we show a deductive reasoning process for the Black-Scholes equation as the concrete example, as many statistical formulas are refereed to derive the Black-Scholes equation.

Keywords: Formula database, Deductive reasoning, Knowledge visualization, Statistical theorem, central limit theorem, Accuracy of estimates, Black-Scholes equation.

1 Introduction

In the paper, we shall discuss math formula databases and its knowledge visualization. Many researchers have been researching mathematical knowledge bases [1], [2], [3], [4]. In the previous paper [5], we pointed out that the existing formula databases lacked excellent user interface and described that a deductive reasoning process can be an effective user interface for a large-scale knowledge-base, using a national income determination problem.

In general, users of formula databases may be a human or a machine; many automatic proof generators have been developed and then the user is a machine, not a human. There strict unique labeling and definition of variables are required. In our

A. Madaan, S. Kikuchi, and S. Bhalla (Eds.): DNIS 2014, LNCS 8381, pp. 132–143, 2014.

research, our target is a human student. The authors are lecturers of business mathematics and we have made many teaching materials. For our lectures, we have developed a web-based education system called "web:VisualEconoMath". The system includes knowledge bases and a lot of teaching materials [6-12]. The purpose of the system is to have our students solve mathematical economics problems. In solving word problems in mathematical economics, such as national income determination problems and various financial problems, two different knowledge bases are required: a database of math formulas and a database of economics theory. Solving a word problem in mathematical economics is nothing more or less than conducting a process of deductive reasoning to find the unknown of the problem. The reasoning is just deduction, although there are some other reasoning methods such induction [13].

In the paper, we discuss the visualization concerning the math formula database. Especially, we shall in the paper describe our approach in the statistical field. For example, we discuss visualization of the central limit theorem. As a concrete example of statistical visualization, we focus on the Black-Scholes model. The Black-Scholes equation is an equation for pricing an options [14-16] and is the most fundamental theory in the financial mathematics.

The derivation of the Black-Scholes equation is a hard word math problem because the derivation requires knowledge of many statistical formulas and theorems. The paper shows illustration of the reasoning process as that can be a good example to show our knowledge visualization approach. In the paper, we shall describe the two things (1) Visualization of reasoning processes can be used as an excellent user interface for a large-scale math knowledge base, and (2) In a statistical field, simulations and its visualization are helpful materials for learners to understand the formulas and theorems.   To implement the (2), at least for statistical teaching materials, the interactive programs for simulation and visualization should also be stored as a part of knowledge as well as video teaching materials.

In Section 2, we illustrate as a concrete example that the reasoning process of the Black-Scholes model is expressed and stored as a part of our math formula knowledge base. Section 3 describes visualization of statistical theorems such as the central limit theorem and accuracy of estimates. Then in Section 4 we discuss the visualization of statistical theorems as a user interface and as a teaching resource. Finally we conclude the paper and describe our future work.

2   Solution Plan Graph of Black-Scholes Problem

In this section, we illustrate as a concrete example that the reasoning process of the Black-Scholes model is expressed and stored as a part of our math formula knowledge base.

First, we will briefly describe a solution plan graph. We call a visual graph of a reasoning process a solution plan graph. In general, a word problem consists of the given data and the unknowns. Given a word problem in mathematical economics, students must first identify the given data and the unknowns. We believe that one of the most significant objectives of mathematics lectures is to cultivate students' reasoning skills. Therefore, in our classes we use our original heuristic method called the Inference Engine Method[17]. We have used this Inference Engine Method since 2007.

We shall explain this Inference Engine Method using the Black-Scholes model as an example problem, which is shown in Fig. 1. This is a finance math problem which finds the expected value of the option. The Black-Scholes (BS) equation for pricing option is one of the most powerful innovations in finance. The formula is widely used both to price options and as a conceptual framework for analyzing complex securities[14-16]. In finance, an option is a contract which gives the buyer (the owner) the right, but not the obligation, to buy or sell an underlying asset or instrument at a specified strike price on or before a specified date. The option profit can be expressed as *max(0, S-K)* where S represents a stock price and K represents the strike price.

The derivation of the Black-Scholes equation is a hard word math problem because the derivation requires knowledge of many statistical formulas and theorems. The paper shows the solution plan graph because the reasoning process is a suitable example to illustrate our knowledge visualization approach. We think that because the derivation problem is tough, the effectiveness of reasoning can be appealed.

The deduction is long. Therefore, the solution plan graph is divided to three parts (Figure 1 to 3). The given data of the problem are a stock price S's price-earnings ratio equation (precisely a stochastic differentiation equation), a drift coefficient μ, volatility σ, the initial stock price S_0, a strike price K and so on. The unknown is the expected value of a call option's profit. The reasoning uses many formulas. The main theorems used there are the Ito's Lemma, Quadratic Limited Variation, and the Central Limit Theorem (CLT). The details of the theory derivations are skipped in this paper [14-16].

To solve the problem, the learner (student) may have conducted the reasoning as follows:

- Because the unknown is the expected value, there must be the probability density function (PDF) of the option profit (S-K).
- The PDF of the option profit (S-K) is strongly related to the PDF of a stock price, because K (the strike price) is fixed. Then, we must get the PDF of the stock price S.

This deductive reasoning is "working backwards". To explain this method, Polya quated words of Pappus of Alexandria as follows [1]: *Let us start from what is required and assume what is sought as already found. Let us inquire from what antecedent the desired result could be derived.* We assent to this Polya's idea and we work backwards to solve the BS equation problem.

Suppose that the stock price fluctuation is expressed by Brownian motion $B(t)$. Then, we set up the stochastic differential equation $S/dS = \mu dt + \sigma B(t)$ which is a precondition and used as a part of the given data. The primary feature of the derivation is to solve this equation using Ito's Lemma. As a result, the function S(t) is found which includes the Brownian motion. The Brownian motion follows the normal distribution, which is derived using the CLT. Finally as shown in Figure 1, we find the PDF of S has a log normal distribution.

The deduction is so long. Therefore many learners (students) cannot grasp the whole structure. However the solution plan graph gives them the bird view of the solution. The solution plan graph is helpful.

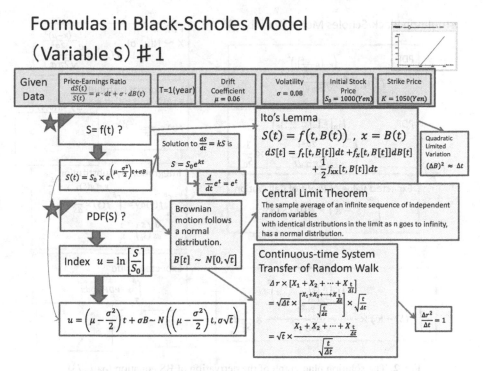

Fig. 1. The solution plan graph of the derivation of BS equation (part 1/3)

3   Visualization of Statistical Formulas

In this section, we focus on statistical formulas and theorems. In the knowledge bases, there are many statistical theorems/formulas. Among them, the most important one is the central limit theorem (CLT). We shall show our developed visual materials for the CLT. In addition, the theorem of 95 percent confidence intervals as an application of the CLT will be also illustrated.

First, we shall describe our recommendation description of the central limit theorem [18], which means there are various descriptions/expressions of the theorem and some may not be easy for students to understand; our target user is our student in business mathematics classes, not student in a mathematics department. The following is one that we are using in our statistics classes.

The central limit theorem states:

Let X1, X2, ... be an infinite sequence of independent random variables with identical distributions (Each X has mean μ and variance σ^2.) Then let $\bar{x} = \frac{X1+X2+\cdots+Xn}{n}$. Then E($\bar{x}$)=$\mu$ and Var($\bar{x}$)= σ^2/n. The central limit theorem says that, in the limit as n goes to infinity, $\bar{x}$ has a normal distribution.

Formulas in Black-Scholes Model (Variable S) #2

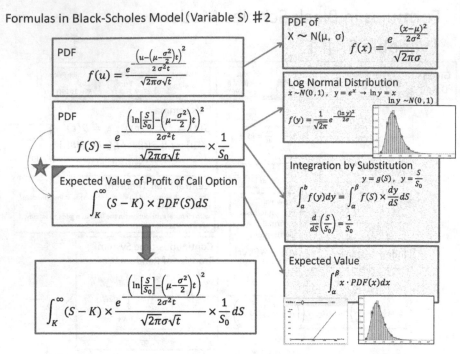

Fig. 2. The solution plan graph of the derivation of BS equation (part 2/3)

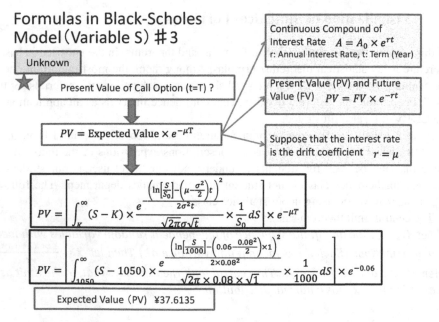

Fig. 3. The solution plan graph of the derivation of BS equation (part 3/3)

The sequence of random variables is independent and identically distributed (abbreviated as i.i.d.). Namely, each random variable has the same probability distribution as the others and all are mutually independent. When we teach the CLT, we had better use a non-normal population variable, so that a normal distribution should be appeared finally and dramatically as illustrated in Figure 4 where the population size is 100 and the sequence of random variables is i. i. d.

Our visual materials use the simulation results of the sampling distribution of means with various sample sizes. On the graphics material, the learners (students) can change the sample size n, dragging the slider. With a big sample size n, the distribution gets close to a normal distribution. The sampling distribution of means is very nearly normal for the sample size $n \geq 30$ even when the population is non-normal[19].

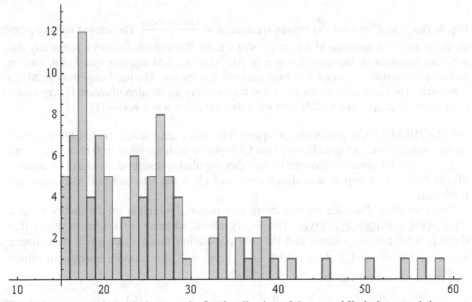

Fig. 4. A non-normal population sample. In visualization of the central limit theorem, it is more effective to use a non-normal population distribution like this.

Some statistical textbooks also adopt simulation results. Compared to them, our simulation materials' feature is its gradual transformation of the probability distribution function. We would like to show learners the effects of the sample size fluctuation. If the sample size n is one, then the probability distribution function of the sample average is similar to the population distribution. Then with a bigger sample size n, the probability distribution function of the sample average gets close to a normal distribution. Looking at the transformation is interesting.

We have published our graphical teaching materials on our web site[1]. The materials are written in Wolfram CDF[2] which is a free mathematical software offered by Wolfram,

---

[1] Yukari Shirota: Graphical Teaching Materials for Statistics,
http://www-cc.gakushuin.ac.jp/~20010570/mathABC/
[2] Wolfram: Wolfram CDF player (Interactive Computable Document Format) site,
http://www.wolfram.com/cdf-player/

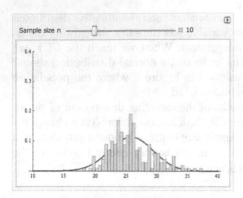 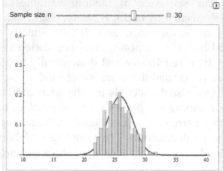

Fig. 5. The simulation results of sample averages, $\bar{x} = \frac{X1+X2+\cdots+Xn}{n}$. The left one shows a probability distribution functions of the sample size n is 10. The right one shows a probability distribution functions of the sample size n is 30. When we add together many i.i.d. random variables, the resulting average will have a normal distribution. The number of trials is 200 on both trials. The simulation results show that the sampling distribution of means is very nearly normal for the sample size $n \geq 30$ even when the population is non-normal[19].

the MATHEMATICA production company. The users can execute every CDF materials on our web site without installation of the CDF player with the latest web browsers. In our web site the CLT simulation materials for other population distributions like (1) a uniform distribution, (2) an exponential distribution, and (3) a discrete uniform distribution are published.

Next we shall illustrate another interactive material which is for the theory about a <u>95 percent confidence interval</u>. The theory about accuracy of estimates states that there is a 95 percent chance that the true population mean value μ will be between $\overline{X_n}(\omega) - 1.96\frac{\sigma}{\sqrt{n}}$ and $\overline{X_n}(\omega) + 1.96\frac{\sigma}{\sqrt{n}}$. There, $\overline{X_n}$, the sample average of which size is n is a random variable.

$$P\left(\left\{\omega : \overline{X_n}(\omega) - 1.96\frac{\sigma}{\sqrt{n}} < \mu < \overline{X_n}(\omega) + 1.96\frac{\sigma}{\sqrt{n}}\right\}\right) = 0.95$$

In our lectures, first we show them the real illustration for the explanation to make them have an image, and then, we make students use the visual tool shown in Figure 6. There the population mean value is invisible. However, with more trials of spraying ink, the mark color gets deeper so that the normal distribution of the true mean can get visible. In the metaphor, one spraying of the ink corresponds to one trial, the radius of a sprayed circle mark corresponds to the half of the confidence interval. The skill of spraying ink to the invisible mean value corresponds to the confidence level such as 95 percent.

In our math classes, the interactive material greatly helps students understand the theory. After that, using another tool shown in Figure 7, we have students count the probability the interval contains the true population mean. There simulation generates the 95 percent confidence intervals on the graphics. The black dot on the interval line corresponds to the sample average. In this case, we set the confidence level to be 0.95, so the probability the interval contains the true mean gets close to 0.95 if the sample size $n \geq 30$.

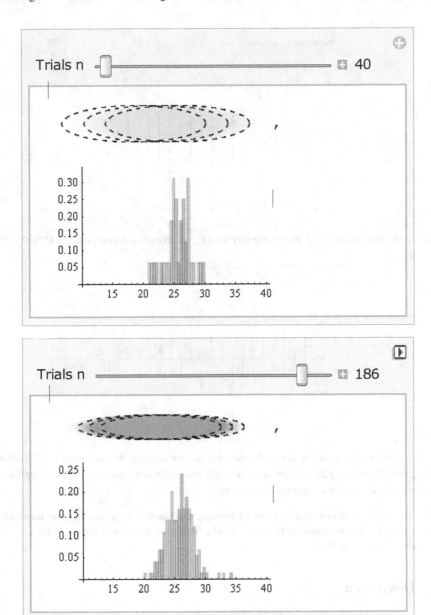

Fig. 6. A graphics material to explain the confidence intervals. The set of red circles correspond to the set of the ink marks sprayed. The centre of a circle is corresponding to a sample average value of one trial. With more trials, the mark colour gets deeper. The upper one is a 40 time trial. The bottom one is a 186 time trial. With more trials, the resulting distribution of the sample averages gets close to a normal distribution. The dot lined circles are the last three sprayed marks. Sometimes, the centre is largely shifted from other circles.

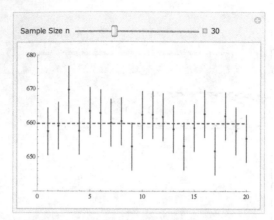

(a)   The sample size is 30. The probability the interval contains the true mean is 18/20=0.9.

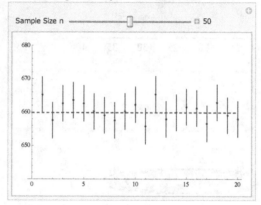

(b)   The sample size is 50. The probability the interval contains the true mean is 19/20=0.95. By using the tool, counting the number of the intervals that contains the true mean, the students get to understand the concept of the accuracy of estimates.

Fig. 7. The visual material to count the probability the interval contains the true population mean. Although the true mean in the case is 660, firstly the true value line is hidden so that students would estimate that.

4   Discussion

In the section, we discuss the construction way of knowledge bases for statistical formulas/theorems.

Our points of the paper are (1) Visualization of reasoning processes can be used as an excellent user interface for a large-scale math knowledge base, and (2) In a statistics field, simulations and its visualization are helpful materials for learners to understand the formulas and theorems.   At least for statistical teaching materials, the interactive programs for simulation and visualization should also be stored as a part of knowledge as well as video teaching materials.

Imagine how a human user uses a formula database. There is no user who would study from one end to another formula thoroughly. They use the formula databases to solve a given math problem. It is limited number of formulas which they are interested in. Therefore, the good interface for the formula database is a tool by which the user can solve the math problem retrieving required formulas. To solve a business word problem is to conduct a deduction. So we believe that a deductive reasoning process can be a good interface to access the formula database.

In the paper, we proved that the reasoning heuristics could be applied to a very complicated word math problem like the Black-Sholes model derivation. The reasoning approach was effective. The used formulas/theorems are well-organized through the solution plan graph.

In July 2013, the author Shirota taught the Black-Sholes model derivation with the visual materials of this paper in Interdisciplinary Faculty of Science and Engineering, Shimane University. The attendants are about 20 under graduate students, graduate students and lecturers. They had no experiences of being given financial math lectures. However, after one and half hours lecture, almost persons could understand the outline of the reasoning. From the informal talk following the lecture, some graduate students stated that he could understand the Black-Sholes model derivation and that although he had already understood the central limit theorem so well, its visualization/simulation was impressive and he would like to use the presentation methods. In addition, the derivation was so long, the solution plan graph helped them understand the whole framework; it showed clearly the given data and the unknown of the problem.

We think that statistical teaching materials should take a different heuristic approach from other math fields such as algebra. In education of statistics and probability, simulation tools should be used. That is our opinion. We believe that real trials such as tossing a coin are very effective so that students can understand definitions and formulas. However, in our experiences, it was too hard to toss a coin more than 200 times. Instead of such a real action, a simulation tool is available. Lecturers of statistics should use simulation tools to have students feel the probability density function.

We think that in a statistics field, simulations and its visualization are helpful materials. It is however difficult to store an interactive program in the knowledge base, compared to text data, math expression data, image data, and video data. The maintenance cost for the interactive programs is higher and the lifetime of the program is shorter. However considering the effectiveness of the programs in the education, we should store the simulation/visualization programs like CDF so that users can teach statistics interactively and visually.

5   Conclusion

In this paper, we have focused on the visualization of statistical theorems and the reasoning process of a word math problem which uses many statistical theorems. As the sample word math, we adopted the Black-Sholes equation derivation. The derivation uses many theorems such as the central limit theorem and the reasoning is complicated. However, showing the reasoning process, it is easier for students to understand that. Through the visualization of the deduction, we proved that visualization of reasoning processes could be used as an excellent user interface for a large-scale math knowledge base. That is our approach and we proposed the reasoning process called solution plan graph as the knowledge base user interface.

Especially in the field of statistics, simulations and its visualization are helpful materials for learners to understand the formulas and theorems. At least for statistical teaching materials, the interactive programs for simulation and visualization should also be stored as a part of knowledge as well as video teaching materials. Considering the effectiveness of the programs in the education, we should store the simulation/visualization programs like CDF in the knowledge bases so that users can teach statistics interactively and visually. We will continually enhance our formula databases and education system.

Acknowledgment. This research is supported in part by Gakushuin University Research Institute of Economics and Management.

References

1. Kohlhase, M., Franke, A.: MBase: Representing Knowledge and Context for the Integration of Mathematical Software Systems. Journal of Symbolic Computation; Special Issue on the Integration of Computer Algebra and Deduction Systems 32(4), 365–402 (2001)
2. Jeschke, S., Wilke, M., Natho, N., Pfeiffer, O.: KEA - a Mathematical Knowledge Management System combining Web 2.0 with Semantic Web Technologies. In: HICSS 2009, pp. 1–9 (2009)
3. W. Research. MathWorld, http://mathworld.wolfram.com
4. Yokoi, K., Nghiem, M.-Q., Matsubayashi, Y., Aizawa, A.: Contextual Analysis of Mathematical Expressions for Advanced Mathematical Search. In: Prof. of 12th International Conference on Intelligent Text Processing and Comptational Linguistics (CICLing 2011), Tokyo, Japan, February 20-26, pp. 81–86 (2011)
5. Shirota, Y., Hashimoto, T., Stanworth, P.: Knowledge Visualization of the Deductive Reasoning for Word Problems in Mathematical Economics. In: Madaan, A., Kikuchi, S., Bhalla, S. (eds.) DNIS 2013. LNCS, vol. 7813, pp. 117–131. Springer, Heidelberg (2013)
6. Hashimoto, T., Shirota, Y.: Web Publication of Visual Teaching Materials for Business Mathematics. In: Proc. of 2nd Uncertainty Reasoning and Knowledge Engineering (URKE2012), Jakarta, August 14-15, pp. 1–4 (2012)
7. Shirota, Y., Hashimoto, T., Kuboyama, T.: A Concept Model for Solving Bond Mathematics Problems. In: Henno, J., Kiyoki, Y., Tokuda, T., Jaakkola, H., Yoshida, N. (eds.) Frontiers in Artificial Intelligence and Applications, Information Modelling and Knowledge Bases XXIII, vol. 237, pp. 271–286. IOS Press (2012)
8. Shirota, Y., Hashimoto, T.: 10 Graphics for Economics Mathematics Part 2 (2011), http://www-cc.gakushuin.ac.jp/~20010570/private/MAXIMA/part2/
9. Shirota, Y., Hashimoto, T.: Bond Mathematics by Graphics - graphics 10 (2011), http://www-cc.gakushuin.ac.jp/~20010570/private/MAXIMA/

10. Shirota, Y., Hashimoto, T.: Visual Instruction Methods for Bond Mathematics Education. In: Proc. of an Annual Meeting of Japan Society of Business Mathematics, Osaka, June 5-6, pp. 63–68 (2010)
11. Shirota, Y., Hashimoto, T.: Animation Teaching Materials for Explaining Recurrence Formula to FInd the Bond Price with the Spot Rate. Journal of Japan Society of Business Mathematics 33, 57–69 (2012)
12. Shirota, Y., Hashimoto, T.: Web Publication of Three-Dimensional Animation Materials for Business Mathematics - 10 Graphics for Economics Mathematics Part2 -. Gakushuin GEM Bulletion 26, 13–22 (2012)
13. Betsur, N.C.: Reasoning Strategies in Mathematics. Anmol Publications Pvt. Ltd. (2006)
14. Brown, A., Mehrling, P.: Fischer Black and the Revolutionary Idea of Finance Wiley (2011)
15. Chriss, N.: Black Scholes and Beyond: Option Pricing Models. McGraw-Hill (1996)
16. Benning, S.: Financial Modeling. The MIT Press (2008)
17. Shirota, Y.: Knowledge Base Construction for Economic Mathematics. Discussion Paper Series, Gakushuin University Research Institute of Economics and Management 5, 1–14 (2006)
18. Downing, D., Clark, J.: Statistics The Easy Way: Barron's (1989)
19. Spiegel, M.R., Stephens, L.J.: Theory and Problems of STATISTICS, 3rd edn. Scgaum's Outline Series. McGRAW-Hill (1988)

Database of Differential Cross Sections for Hydrogen Ionization by Proton Impact

Lukáš Pichl[1] and Daiji Kato[2]

[1] International Christian University
Osawa 3-10-2, Mitaka, Tokyo, 181-8585 Japan
lukas@icu.ac.jp
http://www.icu.ac.jp/

[2] National Institute for Fusion Science
322-6 Oroshi-cho, Toki, Gifu, 509-5292 Japan
kato.daiji@nifs.ac.jp
http://www.nifs.ac.jp/

Abstract. Most of the atomic collision databases in the research data centers are only two-dimensional, typically with the value of collision energy as the abscissa and the cross section value of the process as the ordinate, possibly with up to a few discrete quantum number labels. This design paradigm dating from decades ago, when experiment and theory produced data exclusively in this limited format, has survived to the present day. Meanwhile, multi-dimensional datasets resolved with energy of the process fragments, scattering angles and other variables became customary in the experiment; also recent work in theory often produces highly resolved differential datasets. Nevertheless, the development of new multi-dimensional databases is costly and delayed for most research institutes in the field of atomic physics at present. Scientists are forced to produce two-dimensional projected or integrated datasets to fit the online database format or resign at the task of data upload and sharing. To address this problem, we have developed an online database that supports multi-dimensional cross section data and provides graphical interface for 2D and 3D data plots. The database system is an open-source free-software highly extendible solution organized in modules that can be further reused in an online database builder software.

1 Introduction

A comprehensive and affordable access to information is a fundament of knowledge sharing and research development in science; in the field of atomic physics, researchers share collision process data expressed in terms of the cross section, i.e. the hypothetical area surrounding the target particle that must be hit by the projectile in order for a particular reaction or excitation process to take place. The cross section represents a complex functionality that depends on many parameters of the collision process and can be resolved in terms of the the parameters of the collision product states, both discrete and continuous. Thus there

A. Madaan, S. Kikuchi, and S. Bhalla (Eds.): DNIS 2014, LNCS 8381, pp. 144–151, 2014.

is a need for computing or measuring multi-dimensional cross-section data and their efficient sharing in the atomic and molecular collision process databases. In spite of this significance, information systems used at present by the governments and research institutes are substantially incompatible and rigid, which illustrates how limited database design and insufficient human communication between the research scientist and software developer sides affect the usability of the resulting database systems. We are concerned here with the databases in the atomic and molecular collision physics field as these relate to the simulation of complex processes of energy exchange and material sputtering in the edge plasma at the divertor region of a thermonuclear reactor [1]. The range of applications of the collision cross-section datasets is however substantially broader, and includes atmospheric science, astrophysical processes, radiation therapy in medicine, or particle beam etching in semiconductor industry.

As a case study, we report here the current major online database systems in the National Institute for Fusion Science, Gifu. These are called AMDIS, CHART, SPUTY, BACKS and store the cross sections of collisional excitation, ionization, recombination, charge transfer, sputtering and backscattering data. None of these database systems supports multidimensional datasets. Using our newly computed cross-section data for the process of ionization of atomic hydrogen by impacting protons, which is resolved not only with the energy of the projectile but also the energy of the ejected electron, the established database systems were unable to store the new multi-dimensional dataset, and we have therefore decided to build a prototype database from the scratch. The design specification is rather straightforward: to enable the view of all projections of the multi-dimensional data, which should be integrated with 2D and 3D data plots on demand. The design is object orientated, organized in modules, which allow the integration of the components into an online database builder, under which users can design, implement, initialize and search database on their own, all with the access to general search and data visualization functions.

The paper is organized as follows. Section 2 explains the database design and fundamental requirements for the implementation. The actual system in the production stage is reviewed in Section 3. Concluding remarks close the paper in Section 4.

2   Cross-Section Database

The term cross section in physics is used to indicate the size of a two-dimensional area around a target system that the projectile must hit in order for a certain collisional process to occur. This is a mathematical abstraction, nevertheless: the collision processes at atomic and molecular level are probabilistic phenomena in quantum physics that may take place at any distance between the target and projectile with (although typically very low) nonzero probability. Hence the notion of the cross-section represents an equivalent area, in which the given process can be considered to occur with probability 1, while being forbidden outside. Cross-section data on collisions in fusion edge plasma are typically obtained in experiments (where the above definition has direct implications on the apparatus design) and in theoretical calculations (typically by transforming process probabilities into process cross sections). The units most frequently used for the cross section are atomic units, a_0^2, or the orders of 10^{-16}cm^2. The cross section as a physical quantity depends on a number of physical parameters of the collision process, e.g. the collision energy, quantum numbers of the target molecule etc.

Here we consider the following ionization process,

$$p(E) + \text{H}(p, e) \rightarrow p + p + e(\epsilon, l, m), \tag{1}$$

where the target Hydrogen atom (a system of proton p and electron e) is initially in the ground state. The bound electron is then ionized with an excess energy ϵ and angular momentum numbers l and m. Hence the fully resolved cross section data depend on E, ϵ, l and m, denoted as $\sigma(E; \epsilon, l, m)$. The partial cross sections are integrated over ϵ, and finally the total cross section is obtained by summing up over l and m contributions, $\sigma(E)$. The numerical data for our database were calculated and published by Pichl et. al. [2].

2.1 Database Design

An efficient database design in the community of plasma simulation scientists requires an online access, bibliography information, numerical data available in multidimensional cubes and hyperplane cuts for lower data dimensions, a data search form, and the immediate availability of 2D and 3D plots for use in research papers. A minimal system to test these requirement must therefore have a data structure of the type $f(x_i, y_j; +\text{discrete labels})$, such as the cross section for the process in Eq. (1).

Database Structure. Although this work implements a single database (cross sections for hydrogen ionization by proton impact), the software modules are designed for alternative use as building blocks of a general database builder application. That is why we adopt the object-oriented programming approach for the design of our database.

Figure 1 shows a general scheme of the cross-section database that allows to add functions to the system with recycling the code by the virtue of object-oriented program. The top page links to the search and visualization forms as labeled by the process quantum numbers. As for the hierarchy of object classes, PgSelect class retrieves cross-section data from connections to the PostgreSQL database management system. GraphPgSelect class derives from PgSelect class, and it displays graphs by using the numerical cross-section data. GraphImage 3D, GraphImage E and GraphImage classes derive from GraphPgSelect class, and they display peculiar graphs using functions of GraphPgSelect.

GD Library. For practical applications, it is important that the database interface allows users to copy numerical data into clipboard in suitable formats, and also offers graphical files. We adopted GD for the dynamic generation of graphs.

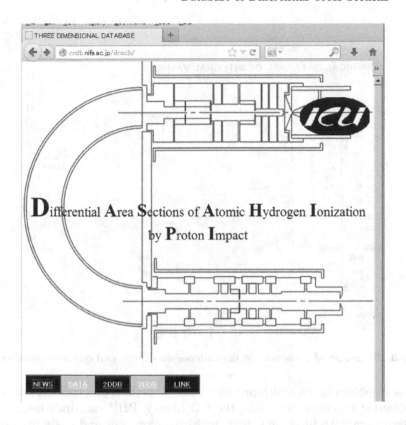

Fig. 1. Top navigation page for the database of differential cross sections

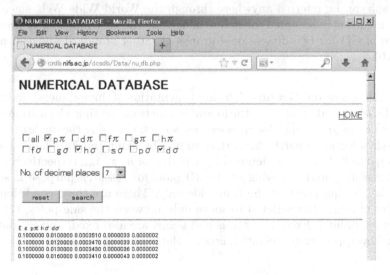

Fig. 2. Retrieval of numerical cross sections for the collision proces

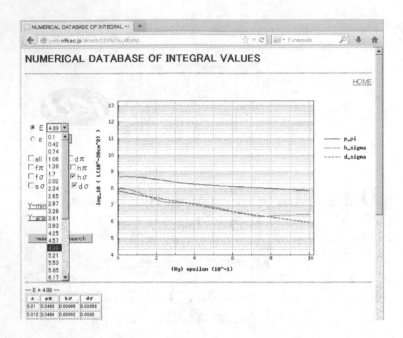

Fig. 3. 2D images of cross section dependence on energy and quantum numbers

GD is a graphics library, which provides PHP (hypertext preprocessor) interface with drawing functions. By using the GD library, PHP can draw images and complete them with lines, arcs, text, multiple colors, cut and paste from other images, flood fills, and also write out the result in the format of a PNG file. The graph can be referred anywhere through the World Wide Web, since PNG format is currently accepted for online images by the vast majority of browsers. The use of GD for two-dimensional graphs is rather standard and will not be detailed here.

Three-Dimensional Graphs. The implementation of three-dimensional graphs with GD is designed to allow for the following functions: setting a logarithmic scale within a predesigned range for cross-section values, changing the angle of view on the graph, selection a particular surface to plot by fixing the state labels l and m (numbered as 0, 1, 2, .. and denoted by s, p, d, .. or σ, π, δ,.., respectively).

As for scaling and centering of the 3D plot, for research purpose it suffices to set the viewing point at the front side only. There are three conditions: (1) the point of origin is restricted to move only between the side poles, (2) when the viewing point is fixed, the x-axis and y-axis are also fixed, and (3) data are re-scaled by appropriate weights following the scaling of the axes.

2.2   Implementation

The system described above was implemented by using a custom-built PC as a dedicated server. The hardware specification is as follows: Celeron(R) 2.00 GHz, 484MB RAM and 120GB HDD, which matches the needs of research database users (large data set, limited access rates). The operating system is Linux (Fedora Core 3) running Apache 2.0 HTTP server. The relational database management system is PostgreSQL 7.4.8 with the connecting logic layer for web interface written in PHP 4.3.11. The above operating system, web and database server and programming language are free-software open-source products. The graphical subsystem producing 3D surface plots in PNG format requires only a very light amount of computational resources, and is therefore best suited for use on the web.

Interface. The implementation of the following three database interfaces is required for the process in Eq. 1:

1. Partial cross sections: $\sigma(E; n, l)$ as a function of E for selectable labels n, l. The numerical data should be displayed with ajustable number of valid digits.
2. 2D slices of the 3D cross-section data cube. All values of E and ϵ in the database should appear in popup menus. Either E or ϵ has to be fixed. Both the graphical form and the tab-delimited numerical form (labeled vector) of output are available.
3. 3D set of data. All l, m labels should appear in a checkbox selection area. Numerical data are present in the form of a labeled matrix, graphical output is available in the form of 3D plot, as described above.

3   Result

The database was implemented exactly as specified in the previous section [3]. Here we describe the three interface modules.

3.1   Numerical Data

Figure 2 shows the data-output form for the partial cross-sections in the multi-dimensional database system. It follows the format frequently used for general spreadsheet software, which divides the data by TAB, and the elements by ENTER. It can display one or more data columns at the same time. Decimal alignment is also available.

3.2   Two-Dimensional Graphs

Figure 3 displays the two dimensional graph output form in the multi-dimensional database system. The online form retrieves the data specified by the user from

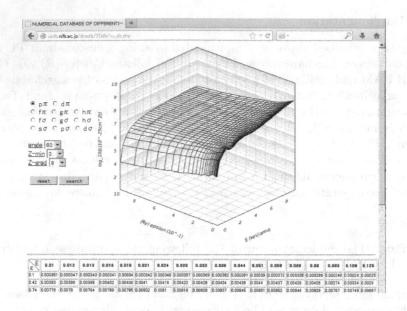

Fig. 4. 3D images of cross section dependendence on the energy of the projectile and the energy of ionized electron in the continuum. Surfaces differ for different quantum numbers.

the multi-dimensional database, and presents them also as a PNG file by using PHP application including the GD library. It can display one or more label sets at the same time. Users can select the minimum value and display magnification of the two-dimensional graph. The dataset is available at the bottom of the interface.

3.3   Three-Dimensional Graphs

Figure 4 shows the three-dimensional graph output form in the multi-dimensional database system. The online form retrieves the data specified by the user from the multi-dimensional database, and presents them also as a PNG file by using PHP application including the GD library. Users can select the minimum value and display magnification of the three-dimensional graph. Pseudo switching gear function is also available and the viewing angle can be set as necessary. The dataset is again available at the bottom of the interface.

4   Conclusion

We have developed a dedicated database system for the cross-section dataset of the hydrogen ionization by proton impact. The solution is based on the open-source free software platform and implemented at a dedicated server at the National Institute for Fusion Science [3]. This complements the commercial databases [4]. Within the framework of an online database builder system, the present solution provides the necessary architecture for storing multi-dimensional datasets and displaying 2D and 3D graphs, which is the first system of this kind in the fusion plasma simulation community. The project of the database builder aims at the development of an online system for database design, implementation, data input, data search and data visualization from one integrated online administration environment, and will be reported in a future paper.

Acknowledgement. The authors would like to thank Mr. Akira Kawano for the implementation and description of the software solution in this database.

References

1. Chen, F.F.: Introduction to Plasma Physics and Controlled Fusion, 2nd edn. Springer, New York (2006)
2. Pichl, L., Zou, S., Kimura, M., Murakami, I., Kato, T.: Total, partial and differential ionization cross sections in proton-hydrogen atom collisions in the energy regeon of 0.1 kev/u - 10kev/u. Journal of Physical and Chemical Reference Data 33, 1031–1058 (2004)
3. Database of the Differential Cross Sections, http://crdb.nifs.ac.jp/dcsdb/ (December 1, 2013)
4. NIFS Databases (December 1, 2013), https://dbshino.nifs.ac.jp/

Handling Domain Specific Document Repositories for Application of Query Languages

Aastha Madaan and Wanming Chu

Database Systems Laboratory,
University of Aizu, Aizu-Wakamatsu Shi, Fukushima, Japan, 965-8580
{madaan.aastha,chuwanming}@gmail.com

Abstract. Domain specific information is increasingly available on the Web in form of document repositories. In specialized domains such as agriculture, bio-medical sciences and health-care, this information is required by various domain experts. Health-care experts such as researchers and practitioners require it during health-care delivery and for educational purposes. These users differ from the Web users and database users. Most of the existing document repositories on the Web have alphabetical and keyword based searches. These are not sufficient for the expert users with precise and complex queries, who require in-depth results within time constraints. Their information needs can be supported by providing user-level schema. Such a schema can support database-style high-level query languages over these repositories. Seeking specialized domain-specific information through queries is gaining importance. In this paper, a model for on-line document repositories is proposed. Queries can be performed with in-depth results. The model can be replicated to similarly structured document repositories in any given domain.

Keywords: on-line domain-specific document repositories, data model, user-level schema, high-level query languages.

1 Introduction

There is a huge volume of domain specific document repositories that are available for scientific purposes [27]. These are often referred by the domain experts in their every-day activities. For example, medical information includes both patient-specific information (EHRs) and knowledge-based information (scientific papers and other literature) [2]. The later is available on the Web through Web document repositories (such as, MedlinePlus ([19]), popular medical literature related publications (PubMed [22], Medline [15]), other primary and secondary resources and EHRs ([8]). Querying these resources is required by the secondary applications such as evidence-based medicine and secondary use of EHRs to improve the quality of care [2]. Medical information is utilized by a variety of end-users with complex requirements. The end-users vary in their background, experience and have variable needs. These users interact in various contexts. A clinician interacts with the patients during clinical-care. He (or she) may need

A. Madaan, S. Kikuchi, and S. Bhalla (Eds.): DNIS 2014, LNCS 8381, pp. 152–167, 2014.

to query the medical literature and other document repositories to assess the plan for the treatments or patient-diagnosis. Limited access to target documents by indexing through a standard search engine is not sufficient. For such queries, the users need a query language and a schema.

A health-care Web-document repository such as an encyclopedia contains documents that represent a large number of common medical conditions, symptoms, diagnostic tests, anatomical and physiological terms, treatments, procedures, prevention topics and other medical terms [13]. These are authentic document collections and offer credible content to the users. For example, MedlinePlus is a free access Web document collection maintained by U.S. National Library of Medicine (NLM) [20]. Over 150 million people from around the world use MedlinePlus each year [30].

According to Kreshmoi survey [14] the end-user requirements vary on the basis of "level of specialty". Practitioners, specialists and researchers are well versed with the medical knowledge and terminologies. These users have precise queries and expect complete results within time limits (almost real-time). The widespread use of WWW has given rise to a range of simple query processors, the search engines. These query a database of semi-structured data (the HTML pages). For example, one can use a search engine to find pages containing a word "tuberculosis". However, it is difficult to obtain only documents in which "tuberculosis" appears in the context of "cause" of "fever" with other symptom as "vomitting".

It provides consumer health information for the patients, their families, doctors and other health-care providers. It brings together information from the United States National Library of Medicine, the National Institutes of Health (NIH), other U.S. government agencies, and other health organizations. Other most commonly used credible health-care document repositories are Health on the Net (Europe), MedlinePlus and Mayo Clinic (North America), Better Health and HealthInsite (Australia) are Web-based health document repositories used by the users are identified by the work of [5]. Web documents of the most commonly cited medical Web document repositories such as the encyclopedias [9], [1], [18], [19] are similarly structured and static in nature. These contain information about medical terms and processes that do not change frequently.

The existing keyword search, menu-based (interface) searches and alphabetical indexes on MedlinePlus and other similar document collections often give redundant results that are of not much use to the medical domain experts. The goal of this study, is to assist the domain experts equipped with domain knowledge but not well-versed to use a query language such as SQL, XQuery with the ability to query the Web document repositories. Understanding of the layout features of these documents and modeling their contents can allow querying on them as attributes of a schema.

1.1 Problem Statement

With the growing need for domain-specific document repositories on the Web, it is generally difficult for the domain experts to search relevant information. They need better, easy-to use query tools to find the relevant results to their precise and complex queries within appropriate time constraints. An earlier work in the area of block-level search, the blocks along with their importance values, have been used to build customized Web search-engines [7]. The approach improves the relevance of search results for record-based or data intensive Web sites, such as, yahoo.com, and amazon.com.

In the case of medical repositories and other domain-specific Web documents, such searches may not be useful, as the context of query is equally important for the user. There is an absence of a well-formed query languages for the Web documents in domains such as life-sciences, biomedical and health-care. Hence, there is a need to enable the domain users in these specialized domains to perform in-depth queries in almost real time. For this the study, proposes to model the specialized document repositories on Web. These have a large number of static Web documents (legacy documents) and the end-users are skilled but lack of interfaces for querying. The proposed model can be replicated to similarly structured document repositories in various specialized domains.

The study is organized as follows. Section 2 describes the details of the model and the underlying algorithm. It also explains the process of querying through query language like XQuery over the transformed user-level schema. Section 3 discusses the application scope of the model. Section 4 describes the related study undertaken. Section 5 summarizes the study and draws conclusions.

2 Proposed Approach

The given section describes the algorithm to model an on-line document repository into a user-level schema. The following definitions help to understand and describe the proposed model.

Definition 1: Web Document Segment Any Web document belonging to a repository of documents can be viewed as a collection of segments. A segment can be defined as a visually distinct and a semantically coherent fragment of a Web document which has a single meaning. A segment may have a label and textual content enclosed within it. The segment label summarizes the enclosed content.

Definition 2: User-Level Schema The terminology-enriched schema in this study refers to the collection of trees. Each tree is a hierarchical structure corresponding to a Web document from the repository. The hierarchical structure is directly mappable to the XML schema. Each of the attributes in the XML schema is defined by a self-explanatory tag which represents a medical terminology or depicts a stage in the patient-diagnosis or preventive care.m For example, causes, disease name, and symptoms represent the tags in the XML schema. These are labels of the segments on the Web document. The content of these segments are enclosed within the corresponding XML attributes.

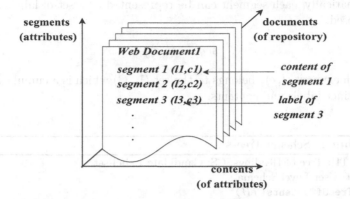

Fig. 1. Representation of the Web document repository as a three-dimensional database

2.1   The Model

In this section, the data model corresponding to a Web document repository is defined with reference to the medical domain. The model is defined for the medical encyclopedia repositories.

A Web document belonging to a medical encyclopedia such as the Medline-Plus medical encyclopedia [19] or University of Maryland's medical encyclopedia [26] can be modeled as a two-dimensional array of segments and their contents. The labels of the segments form the vertical dimension of the array while, the segment contents form the horizontal dimension. All the documents of a document repository can be modeled as a three-dimensional cube (Figure 1). The number of documents form the third dimension of the cube.

For any user query, the segment labels (vertical dimension) describe the predicates and selection values. All the documents (third dimension) are queried for obtaining query results and the segment contents are returned as the results. The queried attributes represent the heading and sub-heading labels of a Web document.

Mathematically, the Web document repository can be defined as a set of documents WD_1, WD_2, ..., WD_n.

$$WDRep = \sum_{i=1}^{n} WD_i$$

Where, n represents the total number of documents in the Web document repository.

Each of these Web documents, WD_i is a two-dimensional array of segments seg_j and their contents which can be represented mathematically as,

$$WD = \sum_{j=1}^{m} seg_j$$

Here, m represents the total number of segments per document of the repository. The number of segments per document is variable.

Mathematically, each segment can be represented as a set of labels and contents enclosed.

$$seg_j = (l, c)_j$$

where, each element $(l, c)_j$ belongs to the set (L, C) which is a cumulative set of all the distinct labels and contents.

Algorithm 1. Schema Tree

Input: TH: Tree of Headings, CS: Candidate Label-set.
Output: User-Level Schema.
`createTreeOfContents(`TH`)`
`createTreeOfSemantics(`$TC,$ CS`)`

2.2   Algorithm

A structural and the semantic analysis is performed for the construction of user-level schema. It uses the knowledge of the pre-established medical processes and stages of the diagnostic processes. This facilitates users to follow a time tested work-flow cycle for data-quality enhancement procedures. Following assumptions can be made about the Web documents in the specialized domains: (i) The HTML parser can successfully extract headings and the subheadings (tags) from the Web documents, (ii) the end-users focus on the main content of the Web document rather than the related content comprising of the meta-data and images and (iii) The segments are assumed to be non-overlapping.

There are three main transformations that are done in order to generate the final hierarchical structure. First the input HTML is transformed to a "tree of headings" (hierarchical structure of HTML heading tags). The root of the tree is the topic of the document and the sub-topics form the intermediate nodes depending on the level of nesting (of the content). Semantic labels or tags are assigned to these nodes, which represent the common stages of diagnostic and preventive care. The resultant hierarchical structure is called "tree of semantics". The "tree of semantics" is parsed into an XML structure. Algorithm 1 gives the two steps of the algorithm.

The method *CreateTreeOfContents()* associates the nodes of the "tree of headings" to the contents enclosed within each of the topics and sub-topics. Next the *CreateTreeOfSemantics()* method, labels the "tree of contents" with labels representative of the domain knowledge.

Structural Analysis. The HTML document is parsed and the heading tags are extracted in form of a hierarchical structure. The Web document is then parsed visually using a set of layout rules (given below) and the headings are marked along a structural curve segregating the content groups. The tree of headings

---

Algorithm 2. Create Tree of Contents

Input: TH: Tree of Headings, LR: 8 Layout rules
Output: Tree of Contents
Perform pre-order traversal of *TH*
Draw a virtual curve in a top-down left right manner on the rendered Web document
for *each node N_i in TH* **do**
    Apply the rules 3-4
    Mark any change in layout (text size, font or color) as *SP*
    Assign the content between 2 consecutive *SPs* to the *SP* above
    Map this node to the corresponding node of *TH*
end

---

and subheadings represents the place-holders for the semantic labels. As shown in Figure 2, any change observed with respect to rules 3-5 (below) is marked as a structure point (SP). On any Web document the content groups are contained between these structure points. The formal algorithm is given as Algorithm 2.

A set of 8 layout (design) rules are identified for recognition of segments on a given Web document similar to the set of rules defined by the VIPS algorithm [3].

Rule 1: Web documents are mostly organized top-down and left-right.

Rule 2: Contents within Web documents are organized in semantic units of information, best suited to user's understanding.

Rule 3: Headings or subheadings within the Web document denote the semantics of the content enclosed within them.

Rule 4: Headings and subheadings of the same group have same text font, size and background color.

Rule 5: Each subheading has different text style (size, font) than the heading within which it is enclosed.

Rule 6: All headings and subheadings have the same orientation.

Rule 7: The components in the Web document may be text/anchors/forms/lists/ tables or images which are recognizable units.

Rule 8: Not all the sub-topic labels are present in every Web document of a repository.

For a given a Web document (WD), the tree of contents (TC) can be mathematically defined as:

$$TC: \sum (hs, c)_i \, , \, i = \{\# \text{ of segments s of the Web document WD}\}$$

Where hs_i is node of the DOM tree (placeholder for the label l_i), representing a heading or subheading in the Web document and c_i represents to the contents under i^{th} label l_i.

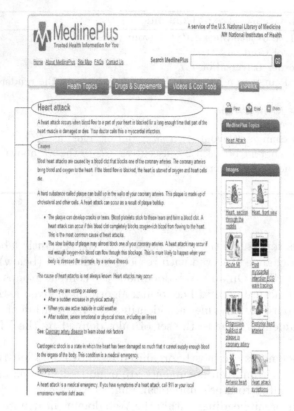

Fig. 2. Structure points (Heart Attack, causes, symptoms) on MedlinePlus "Heart Attack" document

Semantic Analysis. The domain knowledge of any specialized domain is enriched by the resources such as, the dictionaries, thesaurus and concerns databases. These resources form the domain terminologies which further formulate the set of candidate labels for the headings (topics) and the sub-headings (sub-topics) of a Web document. These are available as meta-language components. For example, the medical domain document repository, the medical encyclopedia contains documents which have the segments arranged in a way that follow diagnostic and preventive care processes. The candidate label set for the root level is represented as, CL_{root} = Medical concept/term. Here, the medical concept may refer to a disease name or a laboratory test. The intermediate nodes may refer to the terminologies that describe sub-concepts associated with a disease. For instance, a disease (fever) is a medical term which may have several sub-concepts describing it such as, CL_{fever} = symptoms, causes, exams and tests and treatment. The "tree of headings" is traversed in a pre-order manner and the semantic labels are assigned.

Algorithm 3. Constructing Tree of Semantics

Input: TC: Tree of Contents, *CLnode*
Output: The frequency number (*FreNum$_a$*) node α gets assigned
Perform pre-order traversal of *TC*
for *each node* $n_i \in TC$ **do**
    determineLevel(n_i)
    determine label set corresponding to the level
    AssignLabel(l)
end

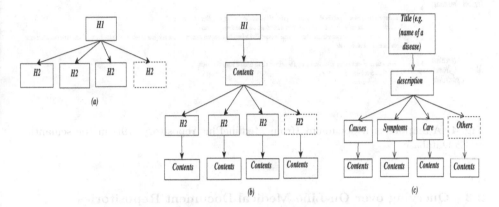

Fig. 3. (a) Tree of Headings similar to a DOM tree (adapted from[11]), (B) Tree of Contents similar to VIPS (adapted [3]), (C) The Semantic Schema Tree corresponding to an example medical encyclopedia Web document

The Tree of Semantics (TS) can be defined as, given the tree of contents (TC) and set of labels (terminologies) CL, from the domain as:

$$TS = \sum (n_i, L)$$

Where n_i is a node in TC and L: TC $\rightarrow$ l maps each node of n_i to a label l where l $\subseteq$ CL, based on the medical terminologies.

The formal algorithm is given as Algorithm 3. The procedure determine_level determines whether the node is a root node or is one of the intermediate nodes of the tree of contents. Figure 3 summarizes the three transformed data structures in form of block hierarchies. The first hierarchy corresponds to the tree of headings, while the second corresponds to the tree of contents and the third represents the semantics-enriched schema tree which is directly transformed to an XML schema.

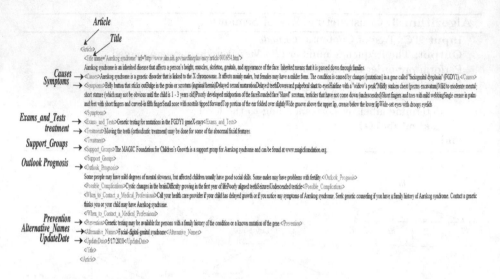

Fig. 4. A Snippet of a document (from MedlinePlus repository [19]) in the semantic XML Database

2.3 Querying over On-Line Medical Document Repositories

Each of the Web documents can be represented as a hierarchy of segments (using described model). The labels of the nodes of these trees represent domain-specific terms and concepts. The schema tree may be visualized as a data graph [25], where the edges represent the attributes and the nodes represent the data. The notion of "path-expression" may be associated with the nodes. Any query can be mapped as a sub-graph of this data graph.

Figure 4 gives the XML database snippet of a MedlinePlus article on "Aarskog syndrome" generated by the above model. As depicted in the figure, each of the XML tag corresponds to a terminology relevant to the health-care process. Similar transformation can be performed on all the articles of different yet similarly structured repositories. Since, the schema is in XML form, the complex task of in-depth querying is reduced to a simpler task of utilizing XQuery query language (or XQBE high-level query language) [6] over the specialized documents repository. Using the semi-structured query language, arbitrary depths of the data graph can be queried by the user. Formally, the semi-structured data can be viewed as an edge-labeled graph with a function.

$$FE: E \rightarrow L,$$

Where L represents the candidate label set corresponding to the domain in consideration forming the edge-labels.

Figure 5 represents the data graph corresponding to a medical encyclopedia document repository. The repository name is the root of the graph and its children are the documents that are classified into various categories (disease, test and surgery). The child nodes show the segments within the Web documents. A user query follows a path from root to leaf i.e. it forms a sub-graph of the schema graph. The hierarchy of the node provides the user the in-depth information as per the user-defined criteria. A query submitted by a user can be viewed as a

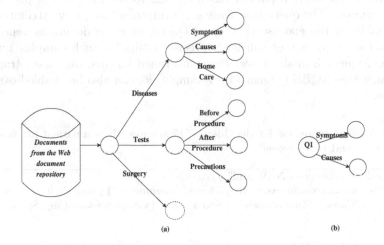

Fig. 5. Representation of a query as a sub-tree of the schema tree (see [25])

sub-tree in the collection of trees corresponding to the document repository. The result of any user query is a sub-tree of the node queried till the leaf node of the sub-tree under that node. If the result of the user query is contained in multiple sub-trees under the root node, then multiple sub-trees are presented to the user. For now, we do not specify any ranking algorithm for the proposed approach, but this would be a taken up as part of future work of this study.

Table 1. Type of Queries addressed by the proposed model with reference to the MedlinePlus medical encyclopedia [19]

S.No.	Query Type	Definition
1	Inter Segment	These are queries with two or more attributes which are subtopic labels in different documents.
2	Intra Segment	These are queries with two or more attributes which are subtopic labels in the same document.
3	Inter Topical	These are queries with two or more attributes which are topic labels or headings in different documents.
4	Topic-Segment	These are queries with two or more attributes, of which at least one is a topic label and at least one is a subtopic label.

Mathematically, the result for a given query on a single or multiple documents of a on-line document repository, can be defined as:

$$Q = \text{return all segments } f\epsilon \sum_{i=1}^{n} WD_i$$

Where f=(l,c) represents the queried concepts, label l corresponding to a segment f defines the queried attribute, and content c contain the selection values.

Table 2 gives a sample XQuery over the user-level schema. The query intends to find the cases where a patient has fever due to the infliction of pneumonia and tuberculosis. The query has "causes", "symptoms" as query attributes (understandable by the end-users). Using XQuery, over the document repository (user-level schema), a user can fetch precise results for such complex queries. Once the XQuery is enabled over the transformed schema, high-level graphical query languages XQBE (XQuery By Example) [6] can also be enabled over the database.

Table 2. XQuery expression for the Query - "Find cases where a patient has fever due to pneumonia and tuberculosis"

```
for $b in doc("AllArticles_NEW.xml")//Article
where ($b/[functx:contains-word($b//Causes, "pneumonia") ] and $b/[functx:contain
s-word ($b//Causes, "tuberculosis") ] and $b/[functx:contains-word($b//Symptoms
, "fever") ] )
return
<Title name=" $b//Title/name "
$b//Title
</Title>
```

Types of Queries. From the point of document view of the users, the approach covers a range of inter and intra topics and segments labels for users to query from the Web document repository. In other words, a user can query single or multiple segments within a single document (intra-segment query). He (or she) can query multiple segments across multiple documents (inter-segment query), he (or she) may query two Web documents labels (inter-topical queries). A user may query a topic and a segment from two or more documents (topic-segment query).

3   Discussions

The proposed model applicable to other similarly structured domain-specific document repositories. In the case of the medical encyclopedia repositories, the other repositories such as University of Maryland, medical encyclopedia [26] and Merck Manual, home health handbook [17] have a similar structure. The segment labels in the Web documents belonging to these are same as the considered encyclopedia. Hence, the proposed algorithm can directly be applied to these

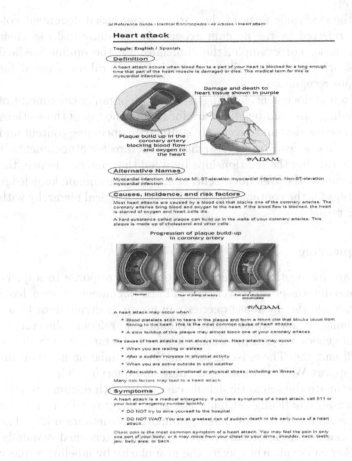

Fig. 6. Heart Attack (Web document) from the University of Maryland Medical Encyclopedia ([26])

document repositories, in-order to make these query-able using a database style query language. Figure 6 depicts the document corresponding to the "heart attack" disease from the University of Maryland, medical encyclopedia. The document contains an array of segments and their corresponding content under the labels of causes, symptoms and treatment. These labels become the query attributes and the results are searched along all the documents containing the attribute or segment label defined by the user within his or her query criteria.

4   Related Work

4.1   Specialized Information Document Repositories

The documents on the Web are well-structured for human readability and comprehension. A program can reliably extract the categorized components information. Specialized domains such as agriculture, law, biomedical, medical domains often comprise of the electronic form of the earlier paper-based document collections. These are referred by the domain experts during daily tasks as their external knowledge base. For example, the clinicians use the on-line medical encyclopedia repositories for patient diagnosis and the general users use it for preliminary symptom recognition.

As in the case of a text document, headings are used to organize the content of the documents. Headings and sub-headings are located at the top of the sections and subsections. These preview and succinctly summarize upcoming content and show subordination. These lead to a hierarchical structure for a document. It plays a role in understanding the relationships between the contents. Hence, the logical (semantic) layout of the Web document, along with the domain knowledge and the structural tags for the content organization form a logical hierarchy with the domain-specific terms.

4.2   Granular Querying

The HTML pages are frequently constructed on the fly in response to a query generated at the user-interface. Moreover, the HTML documents have a fixed level of granularity, while the database queries can group or divide data to arbitrary level of granularity [25]. "Information granulation" refers to the computational processes of generating and presenting levels of abstraction to facilitate problem solving [12] and [29]. The existing information granulation mechanisms do not effectively support Web document searching and querying. They usually fail to accurately estimate the semantic details carried by such documents [27].

An IR system can automatically estimate the granularity requirements of a query using the same approach for document granularity computation [27]. The granularity requirements of an information seeker can be determined manually. An information seeker can explicitly specify the granularity by labeling a query as general or specific (w.r.t. documents required). He or she may use a set of predefined words, such as "review", "introduction", "in-depth" and "specialized" to specify their granularity preferences. In this approach, if all the terms of a document are semantically related, the document is considered as being specific to a particular topic. In the medical domain, the name of a specific medicine or virus is often related to the name of a specific disease. To achieve "information granulation" through domain specific queries, database query-language is required over the on-line domain-specific document repositories.

4.3   Information Needs of the Domain Experts

Several studies on domain expertise have highlighted the differences between experts and novices, including: vocabulary and search expression [23]. The amount of knowledge the domain experts have about a domain is an important determinant of their search behavior. The domain experts can be considered in contrast with Web users, IR users and database users. The Web users search information by browsing the Web documents through urls and are satisfied reading the information on the documents. The database users on the other hand have no schema knowledge but are well equipped and trained with the database query languages SQL and XQuery [28]. The IR users on the other hand use simplistic searches such as keyword search and are generally satisfied reading the documents returned as result for the search criteria. The domain experts such as the medical domain experts, on the other hand are well versed with the schema. They know what they wish to query but do not have easy-to use query tools to formulate their queries over a on-line document repository such as MedlinePlus, WebMD, ADAMS encyclopedia. These users need accurate and precise results for their queries. These users carry out depth-first searches, following deep-trails of information and evaluate the information based on the most sophisticated and varied criteria. Whereas, the novice users concentrate on breadth-first searches and evaluate through overview knowledge [4].

4.4 High-Level Database Query Languages

Database querying is based on the logic of predicate evaluation with a precisely defined answer set for a given query. On the other hand, in an information retrieval approach the ranked results are accepted [21]. Database query languages such as, SQL for relational databases, XQuery for the XML databases have been developed for the skilled database users. These require complete knowledge of query language syntaxes for query formulation. Whereas, the high-level graphical query languages such as, QBE (Query-by-example) for the relational database and XQBE (XQuery by example) for the XML database target both the unskilled and the skilled users. These facilitate query construction. The users are able to quickly form their queries [6]. XQBE allows deep nesting of the XQuery FLOWR expressions [28]. These efficiently meet the query needs of the domain experts because the schema is well-known to them.

5   Summary and Conclusions

The study proposes to model on-line domain-specific document repositories into a user-level schema and enables query-language interfaces over it. It considers the health-care domain for reference. The work emphasizes the need for domain experts to query the on-line document repositories in specialized domains such as health-care, biomedical literature, law and agriculture for everyday tasks. These domain experts have precise and complex query needs which cannot be handled by the existing keyword search. The proposed model is applicable to multiple, similarly structured health-care on-line document repositories. In this study, user-level objects or logically coherent segments of the Web document are extracted to generate a terminology-enriched, user-level database. The study proposed a model to transform an HTML based on-line document repository to an XML schema with semantic tags, understandable by the users. Once, the repository is available in form of an XML schema, query languages such as XQuery can be directly applied over it. This enables the users with powerful and easy-to-use query tools for in-depth querying. The notion of logical and semantically-coherent segments for specialized domain Web documents are used for construction of hierarchical structure. Its orientation is equivalent to the user's view of the document. It is aimed at enhancing query abilities for the domain experts in the specialized domains (such as, the health-care domain).

References

1. A.D.A.M. Medical Encyclopedia (2011), http://www.drugs.com/medical_encyclopedia.html
2. Hanbury, A.: Medical information retrieval: an instance of domain-specific search. In: Proceedings of the 35th International ACM SIGIR Conference on Research and Development in Information Retrieval, Portland, Oregon, USA, pp. 1191–1192 (2012)
3. Cai, D., Yu, S., Wen, J., Ma, W.-Y.: Extracting content structure for web pages based on visual representation. In: Zhou, X., Zhang, Y., Orlowska, M.E. (eds.) APWeb 2003. LNCS, vol. 2642, pp. 406–417. Springer, Heidelberg (2003)
4. Jenkins, C., Corritore, C.L., Wiedenbeck, S.: Patterns of Information Seeking on the Web: A Qualitative Study of Domain Expertise and Web Expertise. IT and Society 1(3), 64–89 (2003)
5. Fisher, D., DeLine, R., Czerwinski, M., Drucker, S.: Interactions with Big Data Analytics. Interactions 19(3), 50–59 (2012)
6. Braga, D., Campi, A., Ceri, S.: XQBE (XQuery By Example): A visual Interface to the Standard XML Query Language. ACM Trans. Database Syst. 30(2), 398–443 (2005)
7. Cai, D., Yu, S., Wen, J.-R., Ma, W.-Y.: Block-based Web Search. In: Proceedings of the 27th Annual International ACM SIGIR Conference on Research and Development in Information Retrieval, Sheffield, United Kingdom, pp. 456–463 (2004)
8. Freire, S.M., Sundvall, E., Karlsson, D., Lambrix, P.: Performance of XML Databases for Epidemiological Queries in Archetype-Based EHRs. In: Proceedings Scandinavian Conference on Health Informatics, vol. 70, pp. 51–57 (2012)
9. Health Illustrated Encyclopedia (2011), http://adam.about.net/encyclopedia/
10. Health Line Medical Encyclopedia (2011), http://www.healthline.com/
11. HTML DOM Tutorial (2011), http://www.w3schools.com/htmldom/default.asp
12. Zadeh, L.A.: Fuzzy sets and information granularity. In: Fuzzy Sets, Fuzzy Logic, and Fuzzy Systems, pp. 433–448 (1996)
13. Laurent, M., Vickers, T.J.: Seeking Health Information On-line: Does Wikipedia Matter? Journal of the American Medical Informatics Association 16, 471–479 (2009)
14. Gschwandtner, M., Kritz, M., Boyer, C.: Requirements of the Health Professional Research, Technical Report D8.1.2, Khresmoi Project (2011)
15. MEDLINE (October 2012), http://www.nlm.nih.gov/bsd/pmresources.html
16. Medical World Search (2011), http://www.mwsearch.com/mwsframetemplate.htm?
17. Merck Manual, Home Health Handbook (August 2013), http://www.merckmanuals.com/home/index.html
18. Middle Georgia Orthopaedics Encyclopedia (2011), http://www.mgo.md/encyclopedia.cfm
19. National Library of Medicine Encyclopedia (October 2012), http://www.nlm.nih.gov/medlineplus/
20. National Library of Medicine, NLM (2011), http://www.nlm.nih.gov/

21. Proceedings of CIDR 2009, Fourth Biennial Conference on Innovative Data Systems Research, Asilomar, CA, USA, January 4-7 (2009)
22. PubMed (2011), http://www.ncbi.nlm.nih.gov/pubmed
23. White, R.W., Dumais, S., Teevan, J.: How Medical Expertise Influences Web Search Interaction. In: Proceedings of the 31st Annual International ACM SIGIR Conference on Research and Development in Information Retrieval, pp. 791–792 (2008)
24. Cohen, S., Kanza, Y., Kogan, Y., Nutt, W., Sagiv, Y., Serebrenik, A.: EquiX-A Search and Query Language for XML. Journal of the American Society for Information Science and Technology 53 (2000)
25. Abiteboul, S., Buneman, P., Suciu, D.: Data on the Web: from Relations to Semistructured Data and XML (2000)
26. University of Maryland, Medical Center, Encyclopedia (2011), http://www.umm.edu/ency/
27. Yan, X., Lau, R.Y.K., Song, D., Li, X., Ma, J.: Toward a Semantic Granularity Model for Domain-specific Information Retrieval. ACM Trans. Inf. Syst. 29(3), 15:1–15:46 (2011)
28. XQuery FLOWR Expressions (2011), http://www.w3schools.com/xquery/xquery_flwor.asp
29. Yao, J.: Granular Computing. In: Proceedings of IEEE International Conference on Information Granulation and Granular Relationships, vol. 1, pp. 326–329 (2005)
30. Zou, J., Le, D., Thoma, G.R.: Combining DOM tree and Geometric Layout Analysis for on-line Medical Journal Article Segmentation. In: Proceedings of the 6th ACM/IEEE-CS Joint Conference on Digital Libraries, Chapel Hill, NC, USA, pp. 119–128 (2006)

Development of eAgromet Prototype to Improve the Performance of Integrated Agromet Advisory Service

P. Krishna Reddy[1], A.V. Trinath[1], M. Kumaraswamy[1], B. Bhaskar Reddy[1],
K. Nagarani[1], D. Raji Reddy[2], G. Sreenivas[2], K. Dakshina Murthy[2],
L.S. Rathore[3], K.K. Singh[3], and N. Chattopadhyay[3]

[1] International Institute of Information Technology-Hyderabad (IIIT-H), India
[2] Acharya NG Ranga Agricultural University, Hyderabad, India
[3] India Meteorological Department, New Delhi, India
pkreddy@iiit.ac.in

Abstract. In several countries, the systems for forecasting weather are being operated to deal with weather and its related factors affecting agricultural production. India meteorological department (IMD) is providing several types of weather forecasts. One of the forecast service is medium range forecast (MRF). As a part of MRF, the expected values of rain fall, temperature, cloud cover, humidity, wind speed and wind direction for next five days are forecasted twice a week by considering district as a unit. Agriculture is markedly affected by weather condition during crop season. IMD in collaboration with Indian Council of Agriculture Research (ICAR) and State Agriculture Universities (SAUs) has set-up about 130 Agro-meteorological Field Units (AMFUs) and each AMFU covers about five districts. Based on MRF, IMD is rendering Integrated Agromet Advisory Service to the farming community of the country in the form of agromet advisory bulletin. The agromet advisory bulletins contain possible risk mitigation measures for the major crops and livestock. Based on the weather forecast, a group of interdisciplinary scientists and agromet scientists at AMFU prepare district-level agromet advisory bulletins. These bulletins are sent to the farmers and other stakeholders of the corresponding district. To ease the process of preparing agromet bulletins, an effort has started to build IT-based agrometeorological advisory system called, eAgromet. In this paper, we explain the concepts of eAgromet and its operation.

Keywords: agromet advice, agromet bulletin, risk management, farm management, extension service, agrometeorology, eAgromet, weather forecast.

1 Introduction

Governments are investing huge budgets and employing advanced computer information systems for weather forecast. The systems for weather forecast are

A. Madaan, S. Kikuchi, and S. Bhalla (Eds.): DNIS 2014, LNCS 8381, pp. 168–188, 2014.
© Springer International Publishing Switzerland 2014

being operated to deal with adverse weather in general to the mankind and agriculture in particular. Over the years, the weather information and forewarning systems are gradually becoming powerful and location specific. India Meteorological Department (IMD) has started weather services for farmers in the year 1945. Currently, IMD is giving the following types of weather forecasts: nowcasting, very short range forecast, short range forecast, medium range forecast (MRF), and long range forecast. As a part of medium range forecast (MRF), from 1st June 2008, IMD has started issuing quantitative district level (612 districts) weather forecast up to five days, twice a week. The MRF comprise of quantitative forecasts for 8 weather parameters viz., rainfall, maximum and minimum temperatures, cloud cover, maximum and minimum relative humidity, wind speed and wind direction. In addition, weekly cumulative rainfall forecast is also provided. IMD generates MRF based on a Multi Model Ensemble Technique using forecast products available from number of models of India and other countries.

Economies of several countries are driven by agriculture and allied activities. Agriculture is significantly affected by weather condition. In India, based on MRF, IMD has launched the scheme "Integrated Agromet Advisory Service (IAAS)" in collaboration with different organisations/institutes to help farming community in carrying out weather-related farm management practices. As a part of this scheme, IMD is issuing agromet bulletins which contain risk management steps for crop and livestock management, based on the weather forecast twice in a week (Tuesday and Friday) up to 5 days. At present, Agromet advisories are being communicated to the farmers of the country through different modes of dissemination systems (Radio, TV, Newspaper, web portals, mobile phones, SMS etc.). IMD is also issuing both national and state-level agromet bulletins. The national- and state-level agromet advisory bulletins are mainly used for planning and resource distribution purposes. The district-level agromet advisory bulletins are prepared and disseminated for the benefit of the farmers of respective districts.

It can be noted that, the crop production systems are influenced by multiple factors such as soil type, crop, variety, location, weather and management practices. To improve crop productivity, farmers need integrated farm advice that consists of advice for crop protection and production problems and appropriate risk mitigation measures based on the weather pattern experienced and experiencing by the crop. The success of agricultural production depends on the degree of preventing/overcoming the ill-effects of crop production factors. Based on the prediction of weather and rainfall patterns, there is a scope to take suitable steps to improve the productivity and reduce the risk. Hence, early warnings based on weather forecasts can help farmers to adjust crop management strategies to minimize the impacts of malevolent climate and maximize the benefits of benevolent climate.

The agromet advisory bulletins provided by IMD are very unique and complements other efforts of improved agriculture technology transfer and agriculture extension methods.

To improve the process of preparing and disseminating agromet bulletins, an effort has been made to investigate the building of an IT-based system to ease the preparation of agromet bulletins, called eAgromet[1]. The main objective is to improve the efficiency of preparation and dissemination of agromet bulletins by exploiting the developments in agriculture, and information and communication technologies. The eAgromet prototype has been developed. In this paper, we explain the main concepts of eAgromet and operation methodology to prepare agromet bulletin.

In the next section, we explain the related work. In section 3, we explain the on going process of agromet advice preparation and dissemination process. In section 4, we explain the basic concepts of eAgromet system. In section 5, we explain the methodology to prepare agromet bulletin with eAgromet system. In section 6, we discuss the performance. The last section contains summary and conclusions.

2   Related Work

In India, the National Center for Medium Range Weather Forecast (NCM-RWF)/IMD, Ministry of Earth Sciences is providing agro-meteorological advisory service based on the medium range weather forecast to the agriculture community on a regular basis. Starting from five units in 1991, NCMRWF/IMD has established about 130 agro-meteorological advisory units and subsequently IMD has taken over and started giving district-wise medium range weather forecast since June 2008. The impact analysis [10] has showed that the weather-based agro-meteorological service is able to reduce the cost of cultivation by 2 to 5 per cent. It was also observed that the advices have improved the yields of various crops. It was also suggested that there is a need to develop a computer-based decision support system, and automate the process of advice preparation and dissemination. The survey [18] regarding current status of agro-meteorological services in South Asia suggests the steps to improve the efficiency of agromet services. Several approaches and issues related to climate prediction and agriculture have been presented in [20].

The forecast of rainfall and temperature distributions can substantially contribute to increased agricultural productivity and farmer livelihood [9] [7]. The Agricultural Production Systems Simulator (APSIM) model developed in Australia has been widely accepted for climate risk management in Agriculture [8]. Efforts are being made to develop decision support systems to help farmers for better crop management based on weather dynamics. In [19], a system called AgClimate has been explained which is a web-based decision support system for minimizing climate risks to agriculture. CERES-Rice and WOFOST models were used to simulate the phenology and yield of low land rice in Telangana region of Andhra Pradesh [14] which helps in taking timely farm management

---

[1] *eAgromet* is the trademark of International Institute of Information Technology-Hyderabad (IIIT-H), Acharya NG Ranga Agricultural University, and India Meteorological Department.

decisions. Efforts have been started to develop content development framework to improve the process of agromet advice preparation [16]. A framework has been proposed for improving practical agricultural skills in [13].

In India, to resolve the crop protection and production related problems, Ministry of Agriculture, Departments of Agriculture, Agricultural Universities, Department of Electronics and Information Technology are making efforts to facilitate the advances in agricultural technology to reach farmers through print and electronic media; organizing seminars and gatherings; Web sites; and telephone. Some of efforts include, farmer portal [1], Kisan Call Centers [4], Digital Green [3], and eSagu [5] [6] [11] [12]. The development of eAgromet system was started in 2010 [15]. In this paper, we will explain the basic concepts of eAgromet and explain the process of agromet bulletin preparation with the usage of eAgromet software.

3   Description of Agromet Advice Service

Weather and climate information play a major role before and during the cropping season and if provided in advance can help farmers apply resources in order to take advantage of favorable conditions and mitigate potential losses in unfavorable ones. The agrometeorological inputs improve agricultural production both in quantity and quality. In an environment of increasing weather and climate variability under climate change, farmers are in greater need of agrometeorological information blended with weather sensitive management advisories before the start of cropping season to support adaptation of agricultural practices. In this context, agrometeorological advisories based on short and medium range weather forecasts become vital to stabilize yields through the management of agro-climatic resources as well as other inputs such as water, fertilizer and pesticides. The IAAS of IMD is intended to contribute to weather information based crop/livestock management strategies and operations dedicated to enhance crop production and food security. The main emphasis of the existing IAAS system is to collect and organize climate/weather, soil and crop information, and to integrate them with the weather forecast information to assist farmers in their management decisions.

3.1   Crop Phenophase, Climatic Normals, Weather Forecast and Agromet Advisory Bulletin

As a part of IAAS service, about 130 agrometeorology field units (AMFUs) operate in different parts of India. Each AMFU covers about five districts. The value added weather prediction values of each district are sent to the corresponding AMFU by the concerned regional meteorological units. The agromet experts at AMFU prepare agromet bulletin for each district.

We explain about the terms *crop phenopahse, climatic normals, weather forecast* and *agromet bulletin.*

Crop Phenopahse: The plant growth can be assessed as a function of completion of a series of phenophases (visible stages of development) that a plant must pass through if it is to grow (and reproduce) successfully. So, the crop growth can be divided into phenophases. Each phenophase duration is specified in number of days. The durations of phenophases may overlap. Based on the sowing date of the crop, we can determine the phenophase of a given crop.

Climatic Normals: Normally, agromet scientists understand the trends of both forecast and observed data and influence of crops and livestock, using the corresponding climatic normals. Climatologists define a climatic normal as the arithmetic average of a climate element such as temperature over a prescribed 30-year interval.

Weather Forecast: The weather forecast consists of the predicted values for the following variables concerning for each district: rain fall (RF), maximum temperature (Tmax), minimum temperature (Tmin), cloud cover (CC), maximum relative humidity (maxRH), minimum relative humidity (minRH), wind speed (WS) and wind direction (WD). The units of RF, temperature, CC, humidity, WS and WD are millimeter (mm), degree centigrade (deg C), octa, percent (%), kilometers per hour (KMPH), and degree (deg) respectively. The sample weather forecast for five days is shown in Figure 1.

Govt. of. India
India Meteorological Department.
Meteorological Centre, Airport colony,
Begumpet airport, Hyderabad 500 016

Value added forecast for next 5 days (Tabular form)
Period: 12-10-2011 TO 16-10-2011. ISSUED ON: 11.10.2011
BULLETIN NO. 81   District: Warangal

DISTRICT : WARANGAL	12/10	13/10	14/10	15/10	16/10
Rainfall (mm)	12	10	9	8	6
Max Temperature (deg C)	33	32	32	33	32
Min Temperature (deg C)	23	23	23	23	23
Total cloud cover (octa)	5	3	6	5	3
Max Relative Humidity (%)	88	87	87	86	86
Min Relative Humidity (%)	76	70	76	70	79
Wind speed (kmph)	3	4	5	3	4
Wind direction (deg)	110	100	140	160	220

Fig. 1. A sample weather forecast

Agromet Bulletin: The agromet advisory bulletin basically contains the information on weather and weather based advisories for crops and livestock. After receiving weather forecast for a given region, the agromet experts prepare the agromet bulletin based on the weather forecast, observed weather values of that region, crop stage and crop status. The agromet advice is prepared in the local language. The components of agromet bulletin are as follows.

- **Agro Meteorological Field Unit (AMFU) details:** Agromet bulletin is being prepared by agromet scientists of AMFU. So, the details of AMFU, unit details, date of forecast and advice valid period are provided.
- **Weather summary text:** The text regarding weather situation about observed weather (till forecast date) and forecast weather of next 5 days (after the forecast received date).
- **Agromet advisory for each crop:** For each major crop of the concerned region, the advice is prepared on the following aspects.
  - **Crop planning:** The advice about the influence of weather forecast on the choice of crop. Information on crop planning, selection of proper sowing/harvesting time etc. and relevant crop husbandry operations are included.
  - **Crop management advice:** After the identification of phenophase, the advice is prepared on how the change in the weather could influence field preparation, sowing/planting, irrigation scheduling, fertilizer application, weed management, pest and disease incidence, their virulence and management operations, harvest and post-harvest handling of crop produce etc.,
  - **Crop management under malevolent weather:** The advisories contain possible mitigation steps for extreme weather events such as extreme temperatures, heavy rains, floods, and strong winds etc. It should contain special steps for taking appropriate measures for saving the crop from malevolent weather are given.
- **Agromet advice for livestock:** The influence of weather forecast on the livestock health (for example, poultry, cattle and buffaloes) and the corresponding management steps are given.

A sample agromet bulletin is shown in Figure 2. It contains the details of AMFU, weather summary text, and weather-based advce for Rice, Cotton, and vegetables (Due to space constraints the advices for other crops have not shown). It also contains the agromet advice for livestock.

3.2   Process of District-Level Agromet Advisory Service

The process of district-level agromet advisory service is divided into four parts as shown in Figure 3.

i **Weather prediction by IMD (input):** The input to the system is the MRF at district level from the IMD, which is being received twice a week on Tuesday and Friday, for five days period. The AMFU receives weather forecast for each district covered by it.

ii **Preparation of agromet advisory bulletin.** To prepare the agromet bulletin, an expert committee consisting of scientists (specialists) from different agriculture disciplines meet on both Tuesday and Friday, and prepares the agro advisories for major crops based on the weather forecast, and existing weather conditions keeping in view the crop status. Agromet advices for

ACHARYA N. G. RANGA AGRICULTURAL UNIVERSITY
Agromet-Cell, Agricultural Research Institute, Rajendranagar, Hyd-
30. **WEATHER BASED AGRO ADVISORIES FOR THE**
WARANGAL DISTRICT FOR THE PERIOD ENDING 15.10.2011
(Till saturday morning)

<u>Bulletin No. XVIII/77/2011</u>   <u>Dt:11.10.2011</u>

During last 24 hrs, light rains occurred. The maximum temperature ranged between 36-37^{0}C and minimum temperature ranged between 22-23^{0}C. As per the forecast received from Meteorological Centre, Hyderabad, light to moderate rains may occur during coming four days. Winds may blow from North West direction with a wind speed of 3 to 5 km/hr. The maximum and minimum temperatures are likely to range between 32-33^{0}C and 22-23^{0}C, respectively.

WEATHER BASED AGRO-ADVISORIES
Rice
• Prevailing weather conditions are congenial for the incidence of brown plant hopper (BPH). Manage the pest by adopting the following measures.
  − Drain out the water from the field
  − Spray Ethofenprox @ 1.5 ml or Acephate @ 1.5 g or Buprofezin @ 1.6 ml per liter of water twice at 7-10 days interval
  − Direct the spray towards the base of the crop
Cotton
• Prevailing weather conditions are congenial for the incidence of aphids. To control,
  − Spray Monocrotophos @ 1.6 ml or Methyl Demeton @ 2ml or Acetamaprid @ 0.2g or Acephate @ 1.5 g per litre of water.
Vegetables
• Prevailing weather conditions are congenial for the incidence of Yellow Vein Mosaic Virus in bhendi. To control the white fly vector, spray Dimethoate @ 2 ml or Acephate @ 1.5 g per litre of water.
Cattle and Sheep
• Prevailing weather conditions are congenial for the occurrence of
  − Hemorrhagic Septicemia, Black Quarter, Foot and mouth disease in cattle,
  − ET and sheep pox in sheep

Principal Scientist (Agromet.)

Fig. 2. A sample agromet bulletin

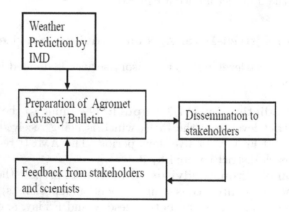

Fig. 3. Preparation of agromet bulletin and dissemination

livestock is also prepared. The agromet bulletin contains agromet advices for major crops and livestock.

iii **Dissemination to stakeholders:** The agromet bulletins are uploaded to web sites of IMD and circulated to press. In addition, the bulletins are disseminated to farmers through private television and radio channels, newspaper, mobile phone/SMS, Internet, farmer portal [1], non-governmental organizations, Kisan (Farmer) Call Centres/Indian Council of Agricultural Research and other related Institutes/state agricultural universities/state extension networks and Krishi Vigyan Kendra (KVKs).

iv **Feedback from stakeholders and scientists:** A group of scientists interact regularly with the farmers to get the feedback which will be used to refine the agromet advice.

4 Concepts of eAgromet

In this section, we explain the motivation and concepts of eAgromet.

4.1 Motivation

Preparation of agromet bulletin is the most important step in IAAS service. Currently, the agromet bulletin is being prepared manually. There are several issues with the existing process of preparation of agromet bulletins. The process is human dependent and consumes significant amount of human effort. There is a possibility of providing generic advice. It requires significant effort and coordination to cover all the crops. It is also difficult to prepare the agromet bulletin by considering several crop-, phenophase-, field-, and weather-specific dynamics at a given location and time.

The main motivation is to develop an IT-based system to ease the preparation of agromet bulletin preparation. For this, a system has to be developed to organize the agromet bulletins in a searchable and re-usable form. The objectives of eAgromet system are formulated as follows.

- The efficiency of the agromet bulletin preparation process should be improved.
- The agromet bulletins should be searchable and reusable.
- The system should be replicable, and
- The system should be simple to understand and operate.

4.2 Concepts of eAgromet

Agromet expert (or subject matter specialist) is a person who analyzes the forecast and observed weather values, and crops' (livestock) condition, and prepares agromet bulletin. It can be noted that the objective is to build an IT-based system to ease the process of agromet bulletin preparation. The system with eAgromet is depicted in Figure 4. The eAgromet system receives weather forecast from IMD through Internet. The agromet scientists access eAgromet and

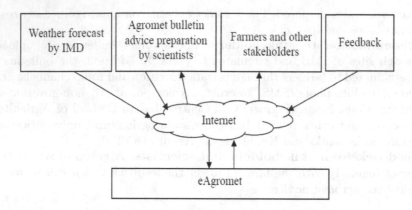

Fig. 4. Agromet Advisory Service with eAgromet

prepare agromet bullentins. The bulletins are delivered to stakeholders. The feedback is also entered into the system.

To develop eAgromet system, several interactive sessions were held with agromet scientists. It was identified that the following steps are being followed by agromet expert for preparing agromet bulletin.

– Comprehend the trend of forecast and observed weather data

  • Analyze the forecast weather data.
  • Analyze the observed weather data till date.
  • Analyze both forecast weather and observed weather data with reference to corresponding climatic normals.

– Prepare agromet advice for each crop by analyzing observed and forecast weather effects on the phenophase of the crops.
– Prepare the agromet bulletin by placing the AMFU details, weather summary, crop-wise agromet advices, and livestock-wise agromet advices.

For agromet scientists, the main issue is to understand the trend of observed and forecast weather, and corresponding influence on the crop and livestock. For this, we introduced the notion of *weather deviation* which will be elaborated as follows: Given the day of forecast, the agromet expert receives forecast values for next five days and observed weather values of the same parameters till the day of forecast. Next, the agromet expert analyzes the observed weather prior to the day of forecast and the forecast weather values and corresponding climatic normals. In other words, the agromet expert tries to comprehend (i) the trend of forecast weather as compared to observed weather (ii) trend of observed weather with reference to the corresponding climatic normals and (iii) trend of forecast weather with reference to the corresponding climatic normals. We call such notion as ***weather deviation***. After comprehending the weather deviation, agromet expert analyzes how the weather deviation affects the crops and livestock, and prepares the corresponding agromet advice.

In addition to weather deviation, we have also developed a component to find similar advice to ease the preparation of agromet advice for a crop and a module to generate weather summary. Overall we have developed three frameworks.

I A framework of weather deviation
II A framework to extract similar advice
III A framework to compute the weather summary

I. A Framework of Weather Deviation

The process of agromet bulletin preparation starts after receiving the weather forecast on some date. Let the notation df be the day of forecast, the notation pd be the past duration which is the number of days preceding to df, and the notation fd be the forecast duration. It was observed that the agromet scientists are analyzing the weather by calculating weather statistics week-wise. The day-wise values are also being examined based on the criticality.

At first, we define the term "weather statistics (ws)". Let d be the duration in days. The weather statistics for a duration d, ws(d) is given in Definition 1.

Definition 1: Weather statistics for a duration d $(ws(d))$: For each day, we receive values for RF, Tmax, Tmin, CC, maxRH, minRH, WS, and WD. The notation $ws(d)$ is the statistics values for RF, Tmin, Tmax, CC, maxRH, minRH, WS, and WD over duration d, i.e., ws(d)=(s(RF), s(Tmin), s(Tmax), s(CC), s(maxRH), s(minRH), s(WS), and s(WD)). Here, the notation $s(x)$ represents statistic parameter value which captures central tendency for the weather variable x for duration d. Note that appropriate function should be employed to compute the statistic parameter value. For Tmax, Tmin, CC, maxRH, minRH, WS, and WD, $s(x)$ is equal to the mean value over duration d, whereas for RF, the $s(x)$ represents the cumulative value over the duration d.

Given df, the change in the weather from pd to fd is captured through the notion of *weather deviation*. The notion of *weather deviation* is the key concept of the eAgromet system. The main issue is to develop a system to ease the comprehension of the vagaries of observed and forecast weather. We first define weather statistics based weather deviation. Next, we enrich the definition by including the aspect of climatic normals of weather variables. We define the final definition by including the notion of *weather categories*.

i **Weather statistics-based weather deviation:** Given df, the weather deviation is indicated by $wd(df)$ which is equal to $<df,\ ws(pd),\ ws(fd)>$, where ws(pd) denotes the weather statistics of pd and $ws(fd)$ denotes the weather statistics of the fd. Note that the df is included in fd. The past duration is divided into 7 days intervals and ws is computed for each interval. By replacing $ws(pd)$ with "ws(-n),.., ws(-2), ws(-1)", the $wd(df)$ is defined as follows. $<df,\ ws(-n),..,\ ws(-2),\ ws(-1),\ ws(fd)>$. Here, ws(-n) indicates the weather summary of n'th previous week with reference to df.

So, the weather deviation captures the trend of observed weather values over several weeks prior to df and the trend of forecast weather values.

ii **Normals-based weather deviation:** It was observed that the agromet scientists compare the weather values with the corresponding climatic normals. So, it is possible to improve the understanding of weather deviation by providing the corresponding weather normals. So, the agromet scientist understands the vagaries of weather and corresponding effects on crops and livestock in a better manner, if the corresponding normal values are shown as a part of weather deviation.

The definition of normals-based weather deviation is as follows. Let $cn(d)$ indicates the climatic normals for duration d. Let the past duration is divided into 7 days intervals. The definition of weather deviation for a given df is as follows: wd(df): $<df, \{ws(-n), cn(-n)\},.., \{ws(-2), cn(-2)\}, \{ws(-1), cn(-1)\}, \{ws(fd), cn(fd)\}>$. Here, $ws(-n)$ indicates the weather statistics of n'th previous week with reference to df and $cn(-n)$ indicates the climate normals of n'th previous week.

iii **Category-based weather deviation:** With *normals-based weather deviation*, the agromet expert can improve the understanding by analyzing observed and forecast weather with the corresponding climatic normals.

We introduce the notion of category to comprehend the weather deviation in a better manner. It was observed that the agromet expert does not give different advice unless a weather value changes to a considerable extent, For example, the agromet expert does not give a different advice for a small change, like 0.2 degree centigrade, in temperature value, or small change, like 2 per cent, in humidity value. Also, having real values for weather parameters for weather statistics and climatic normals, it is difficult to compare weather deviations and build a system to extract similar weather deviations and corresponding agromet advices.

We have improved the definition of weather deviation based on the notion of *category*. For agromet expert, it is relatively easy and quick to grasp the dynamics of change through categories in the weather values over several weeks. Also, the notion of tags ease the comparison of weather deviations by a software system.

Category or tag: For each weather variable, we divide the domain of that weather variable into different classes. Each class is termed as a category or a tag. The tag is a description/name of that class. For example, the domain of temperature can be divided into: VERY COLD, COLD, PLEASANT, HOT, MORE HOT and so on. Similarly, each other weather parameter is divided to different categories.

Note that, as agromet expert is involved in the preparation of agromet bulletin, the tags have been defined such that agromet expert comprehends the weather and then extends how the weather deviation will influence the crops or livestock. Alternatively, the tags can be defined by considering each crop/livestock. It will enable the building of automatic agromet advisory system which will be investigated as a part of future work.

By incorporating the notion of categories, we define the notion of *category-based weather summary* which is as follows:

Category-based weather statistics for a duration d (cws(d)): For each day we receive values for RF, Tmin, Tmax, CC, maxRH, minRH, WS, and WD. Let $c(s(x))$ be the category of the weather statistics of weathr variable x over duration d. For example, let the mean value of Tmin over one week is equal to 16 degree centigrade. So, s(Tmin)=16. Then, if this is classified as COLD, then c(s(Tmin))=COLD. The values of other variables are mapped to the corresponding categories. To design eAgromet prototype system, the tags[2] have been assigned to all weather variables (refer Table 1). Given the weather variable value, the corresponding tag could be obtained from Table 1. The cws(d) is equal to $<c(s(Tmin))$, $c(s(Tmax))$, $c(s(RH))$, $c(s(RF))$, $c(s(R))$, $c(s(WS))$, $c(s(WD))>$.

Based on the definition of category-based weather statistics, we define the final definition of weather deviation as follows. Normally, given the weather forecast, the wd helps to comprehend the dynamics of forecast weather and observed weather of one or several past weeks.

Definition 2. Weather deviation: Let cn(d) indicated the climatic normals for duration d, $ws(d)$ indicates the weather statistics, $cws(d)$ indicates the category of corresponding weather statistics, $ccn(d)$ indicates the category of climatic normals. The definition of $wd(df)$ for a given df is as follows: $<df$, $\{ws(-n)$, $cws(-n)$, $cn(-n)$, $ccn(-n)\}$,.., $\{ws(-2)$, $cws(-2)$, $cn(-2)$, $ccn(-2)\}$,.., $\{ws(-1)$, $cws(-1)$, $cn(-1)$, $ccn(-1)\}$, $\{ws(fd)$, $cws(fd)$, $cn(fd)$, $ccn(fd)\}>$. Given df, the past duration is divided into 7 days intervals. Also, both ws and cws are computed for each 7 days interval. The notations ws(-n), cws(-n), cn(-n), and ccn(-n) indicates the weather summary, category-based weather summary, climatic normals and category of weather normals of n'th previous week respectively.

So, for each variable, the weather deviation shows the corresponding statistics value, summary tag, normal value and normal tag for several preceding weeks and forecast duration. Based on the age of the crop, the agromet expert can easily comprehend the influence of weather deviation on the crop situation and prepare the agromet advice for the corresponding crop by applying the agro-climatic expertise. The example of weather deviation is shown in Figure 6.

II. A Framework to Extract Similar Advice

In addition to weather deviation, another important concept which is being developed is **the module to extract similar advice**. It is expected that, the eAgromet system is populated with several agromet advices of multiple crops attached with corresponding weather deviations. The data of agromet advices are maintained in the eAgromet system, in the following form: $<crop$ and $location$, wd, $weather$ $summary$, $agromet$ $advice>$. After receiving the weather forecast and observed weather values, the wd is formed. Based on the crop, location, weather summary and wd, the module extracts similar agromet advices and displays as per the rank order. The agromet expert can select the agromet advice and make the modification, if required.

---

[2] The tags have been assigned by the design team to develop eAgromet.

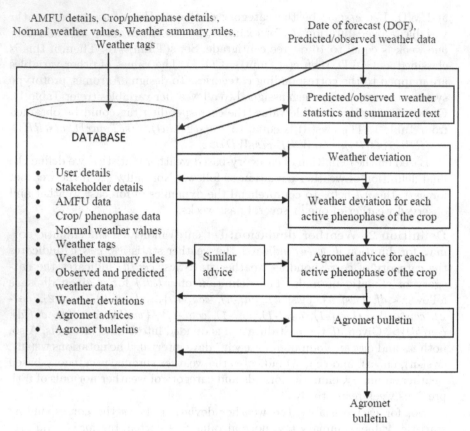

Fig. 5. Steps of agromet bulletin preparation with eAgromet. Here, the arrow indicates the data flow and rectangles indicates the processes. The DATABASE indicates several types of data maintained by the system. The process starts with the entering of forecast weather and observed weather for the region. The system facilitates the preparation of weather statistics and summary, weather deviation, crop-specific agromet advice and agromet bulletin. The agromet bulletin is disseminated to stakeholders.

III. A Framework to Compute the Weather Summary

Agromet bulletin contains weather summary and major crop- and livestock agromet advices. We have already explained how the notion of weather deviation and the method to extract similar advice have been developed to ease the process of agromet advice preparation. In addition, we have also added a module to compute weather summary based on weather summary rules.

Based on the weather summary rules, the system selects the summary sentences to form weather summary. The agromet expert can modify, if necessary. The rules[3] for forecast rain fall are given in Table 2 and the rules for observed rain fall are given in Table 3. The rules for temperature are given in Table 4. The

---

[3] It can be noted that these rules were developed by the design team to build eAgromet prototype. These rules were not standardized by India Meteorological Department.

Table 1. Tags for weather variables. For each weather variable, the tags were assigned which are self explanatory.

Tag Type	Range	Description (Tag)
Rain	0–0	No Rain (NR)
Fall	0.1–2.4	Very Light Rain (VLR)
(mm)	2.5–7.5	Light Rain (LR)
	7.6–35.5	Moderate Rain (MR)
	35.6–124.4	Heavy Rain (HR)
	124.5–244.4	Very Heavy Rain (VHR)
	≥ 244.5	Extremely Heavy Rain (EHR)
Tempe-	≤-5	Freezing (F)
rature	-5– -3	Extreme chill-5 (ECH5)
(°C)	-3– -1	Extreme chill-4 (ECH4)
	-1–1	Extreme chill-3 (ECH3)
	1–3	Extreme chill-2 (ECH2)
	3–5	Extreme chill-1 (ECH1)
	5–7	Very chill (VCH)
	7–9	More Chill (MCH)
	9–11	Chill (CH)
	11–13	Very Cold (VCOLD)
	13–15	More Cold (MCOLD)
	15–17	Cold (COLD)
	17–19	Very Cool (VCOOL)
	19–21	More Cool (MCOOL)
	21–23	Cool (COOL)
	23–25	Cool & Pleasant (CP)
	25–27	Pleasant (P)
	27–29	Warm & Pleasant (WP)
	29–31	Warm (Wa)
	31–33	More Warm (MW)
	33–35	Very Warm (VW)
	35–37	Hot (Ho)
	37–39	More Hot (MH)
	39–41	Very Hot (VH)
	41–43	Extreme Hot1 (EH1)
	43–45	Extreme Hot2 (EH2)
	45–47	Extreme Hot3 (EH3)
	47–49	Extreme Hot4 (EH4)
	49–50	Extreme Hot5 (EH5)
	≥ 51	Unbearable Hot (UH)

Weather Tags

Tag Type	Range	Description (Tag)
Relative-	0–30	Low (L)
Humidity	31–60	Moderate (M)
(%)	61–80	High (H)
	≥ 81	Very high (VHi)
Cloud-	0–2	SKY Clear (SC)
Cover	3–5	Partly Cloudy (PC)
(Okta)	6–7	Mostly Cloudy (MC)
	7–8	Cloudy (C)
Wind-	1 –20	North North East (NNE)
Direction	21–50	North East (NE)
(Degrees)	51–70	East North East (ENE)
	71–90	East (E)
	91–110	East South East (ESE)
	111–140	South East (SE)
	141–160	South South East (SSE)
	161–180	South (S)
	181–200	South South West (SSW)
	201–220	South West (SW)
	221–250	West South West (WSW)
	251–270	West (W)
	271–290	West North West (WNW)
	291–320	North West (NW)
	321–340	North North West (NNW)
	341–360	North (N)
Wind-	0–0.9	Calm (Ca)
Speed	1–5	Light Air (LA)
(kmph)	6–11	Light Breeze (LB)
	12–19	Gentle Breeze (GB)
	20–28	Moderate Breeze (MB)
	29–38	Fresh Breeze (FB)
	39–49	Strong Breeze (SB)
	50–61	Near Gale (NG)
	62–74	Gale (G)
	75–88	Strong Gale (SG)
	89–102	Storm (St)
	103–117	Violent Storm (VS)
	≥ 118	Hurricane (Hu)

Table 2. Weather summary rules for forecast rain fall (RF)

MinRF	MaxRF	RF Summary
0	0	Dry weather may prevail in coming five days.
0.1	2.4	Possibility of light rain at one or two places.
0.5	1	Light rains may occur in coming five days.
1.1	3	Light to moderate rains may occur in coming five days.
7.6	12	Moderate rains may occur in coming five days.
7.6	30	Moderate to rather heavy rains may occur in coming five days.
30	50	Rather heavy rains may occur in coming five days.
50	100	Heavy rains may occur in coming five days.
100	120	Very heavy rains may occur in coming five days.

rules for wind direction and wind speed are given in Table 5. The rules for cloud cover are given in Table 6. The sample weather summary generated is shown in Table 7.

Table 3. Weather summary rules for observed rain fall (RF)

MinRF	MaxRF	RF Summary
0	0	During last week Dry Weather prevailed in the district.
0.1	2.4	Light rain at one or two places.
2.5	7.5	Light rain at some places.
7.6	12	Light rain at most of the places.
20	30	Moderate rainfall.
30	50	Heavy rainfall.
50	100	Rather heavy to very heavy rainfall.
100	120	Very heavy rainfall at many places.

Table 4. Weather summary rules for temperature and relative humidity to the forecast and observed period. Here, '*' indicates the value and 'Deg.C' indicates the degree centigrade.

Weather Parameter	Summary
Temperature Forecast	Next week, the minimum and maximum temperature range between * and * Deg.C.
Temperature Observed	Last week, the minimum and maximum temperature ranged between * and * Deg.C.
Relative Humidity Forecast	The minimum and maximum relative humidity range between * and * %.
Relative Humidity Observed	The minimum and maximum relative humidity ranged between * and * %.

Table 5. Weather summary rules for wind speed (WS) and wind direction (WD). Here '*' indicates the mean value of WS. The rules to generate summary for forecast and observed period are the same, except that we use the word 'blow' in stead of 'blew' for the forecast period.

MinDeg	MaxDeg	Direction Summary
0	50	North-Eastern winds blew (blow) at a speed of * kmph.
51	90	Eastern winds blew (blow) at a speed of * kmph.
91	140	South-Eastern winds blew (blow) at a speed of * kmph.
141	180	Southern winds blew (blow) at a speed of * kmph.
181	220	South-Western winds blew (blow) at a speed of * kmph.
221	270	Western winds blew (blow) at a speed of * kmph.
271	320	North-Western winds blew (blow) at a speed of * kmph.
321	360	Northern winds blew (blow) at a speed of * kmph.

Table 6. Weather summary rules for cloud cover (CC). The rules to generate summary for forecast and observed period are the same, except that we use the word 'may prevail" in stead of 'prevailed" for the forecast period.

MinOkta	MaxOkta	Cloud Cover Summary
0	2	The clear sky weather prevailed (may prevail).
3	5	The partly cloudy weather prevailed (may prevail) The clear sky prevailed.
6	7	The mostly cloudy weather prevailed (may prevail).
7	8	The cloudy weather prevailed (may prevail) The clear sky may prevail.

Table 7. A sample of weather summary computed by eAgromet. It contains observed weather data for 7 days and forecast weather data for 5 days. The summary is generated by the system based on the weather summary rules provided in Table 2, 3, 4, 5.

Date	RF (mm)	Tmax (^{o}C)	Tmin (^{o}C)	CC (Okta)	maxRh (%)	minRh (%)	WS (kmph)	Wd (degrees)
Observed Weather Data								
17/Dec/2013	0	31.5	11.8	0	88	64	1.8	0
18/Dec/2013	0	31.0	12.4	0	91	64	2.2	20
19/Dec/2013	0	30.0	13.0	0	90	71	2.0	0
20/Dec/2013	0	29.8	13.8	0	91	64	1.9	0
21/Dec/2013	0	29.8	10.0	0	86	80	1.8	0
22/Dec/2013	0	28.0	9.0	0	92	85	2.5	20
23/Dec/2013	0	29.0	14.9	0	84	78	4.0	360
Forecast Weather Data								
24/Dec/2013	0	29	14	0	56	31	4	197
25/Dec/2013	0	28	14	0	43	34	5	163
26/Dec/2013	0	27	13	0	44	26	5	64
27/Dec/2013	0	29	15	1	59	30	5	260
28/Dec/2013	0	27	14	0	47	28	6	64
Observed Weather Summary								
During last week Dry Weather prevailed in the district. The minimum and maximum temperature ranged between 9.0-14.9 and 28.0-31.5 Deg.C. The minimum and maximum relative humidity ranged between 64.0-85.0 and 84.0-92.0%. North-Eastern winds blew at a speed of 1.8-4.0 kmph.								
Forecast Weather Summary								
As per the forecast received from IMD, Ahmedabad, Dry weather may prevail in coming five days. The minimum and maximum temperature range between 13.0-15.0 and 27.0-29.0 Deg.C. The minimum and maximum relative humidity range between 43.0-59.0 and 26.0-34.0%. Eastern winds blow at a speed of 4.0-6.0 kmph.								

5   Preparation of Agromet Bulletin with eAgromet

The notions of weather deviation, module to extract similar advice, weather summary text generation are the important concepts in eAgromet. In addition, the eAgromet system contains several types of databases. The process of agromet bulletin preparation with eAgromet system and the corresponding data details are given in Figure 5. Here, the term *region* indicates the area for which weather forecast is received and agromet bulletin is being prepared. Currently, *region* indicates the *district* (administrative unit) in India.

The following are the steps to be followed for preparing agromet bulletins.

i Fill the region and crop details
ii On receiving weather forecast, follow the steps below to prepare the agromet advice.
   (a) Preparation of weather statistics and weather summary text.
   (b) Preparation of weather deviation
   (c) Preparation of agromet advice for each major crop, and major livestock
   (d) Preparation of Agromet bulletin

The details of each step are discussed as follows.

i **Fill the region and crop details.** The following details are entered into the system.
   - **AMFU details:** The name and address of AMFU, details of the districts, details of scientists and stakeholders, and agro-climatic information and user information.
   - **Crop/phenophase data:** Details of crops and the details of corresponding phenophases. Details of livestock are also entered.
   - **Observed weather data:** The observed weather data of the concerned region is being entered.
   - **Normal weather data:** The climatic normals of weather parameters for each district.
   - **Weather tags:** Details of weather categories employed in the prototype are given in Table 1.
   - **Weather summary text rules:** The rules for generating weather summary. The rules developed for the prototype are given in Figure 2, 3, 4, 5

ii(a) **Preparation of weather statistics and weather summary text:** After receiving the weather forecast data and observed weather data for a given region, it is entered into the database. For the forecast data, five day weather statistics is prepared. It is nothing but the mean values of Tmax, Tmin, CC, maxRH, minRH, WS and WD and cumulative value for RF. For the observed data weather statistics for 7 day period preceding to df is prepared. In addition, based on weather summary rules, weather summary is computed and shown to agromet expert for further editing/processing. The sample weather summary produced is shown in Table 7.

ii(b) **Preparation of weather deviation:** The weather deviation is computed as per Definition 2. Through weather deviation, the agromet expert can comprehend (i) weather phenomena prior to df and during forecast period and (ii) the deviation of weather with reference to the corresponding climatic normals. The tags help the agromet expert to comprehend the weather deviation. The weather deviation is stored in the database. A sample weather deviation is shown in Figure 6.

ii(c) **Preparation of agromet advice for each crop, and each livestock:** After forming weather deviation, the agromet expert prepares the agromet advice for each crop and livestock. Given a crop and weather deviation, the *similar advice* tool extracts the similar advices from the repository of agromet advices. By comparing the region, crop, phenophase and weather deviation, the system displays the possible similar advices starting from most similar advice. The agromet expert can select the similar advice and carry out appropriate modification. The prepared advice is stored in the database.

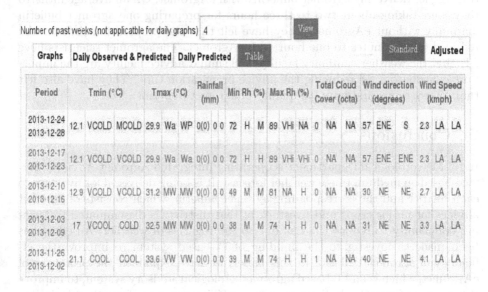

Fig. 6. A sample weather deviation computed by eAgromet system. The notation 'NA' indicates 'not applicable'. It means that the value is not entered. The date of forecast is 24-12-2013. The first column shows the period and the other columns show the weather parameters: Tmin, Tmax, Rainfall, minimum RH, maximum RH, Total cloud Cover, Wind Direction and Wind Speed. The table contains five rows. The first row shows the weather statistics for the forecast period till 28-12-2013. Each of the other four rows show the weather statistics for the observer weather for past four weeks (the number of past weeks can be changed based on the requirement) preceding to the date of forecast. Except rainfall variable, the values for other variables are displayed in three columns as per the following order: <mean value of the weather parameter, weather statistics tag, weather normal tag>. The meaning of tags are given in Table 1. The rainfall values are displayed as the following order: <cumulative value of rainfall (the number of rainy days), normal value of cumulative rainfall, number of normal rainy days>. In addition, tools are provided to visualize the weather through *graphs* focusing on daily values and forecast values.

ii(d) **Preparation of Agromet bulletin:** The agromet bulletin is prepared by combining weather summary text and agromet advices of crops and livestock. The document is stored in the database which will be disseminated to stakeholders. The agromet bulletin for a given region is the combination of advices of the crops in that region. Based on df, the weather deviation is formed and advices for all crops have been entered. Based on df and region, all the advices are combined in an appropriate manner for forming the agroment bulletin.

6   Performance

The eAgromet prototype was developed. About 30 agromet scientists were given training for delivering agromet bulletins with eAgromet. On an average hitherto they were taking about two to three hours for preparing one agromet bulletin manually without eAgromet. They have felt that the preparation time can be reduced to 30 minutes to one hour with eAgromet. The agromet scientists have felt that the weather summary facility, and similar advice facility are the important contributors for reducing the advice preparation time. They have also felt that the system could improve quality of agromet bulletins.

7   Conclusions

India Meteorological Department is operating Integrated Agromet Advisory Service based on medium range weather forecast. It is being delivered for five days to each district and the corresponding agromet bulletin, which consists of agromet advices for major crops and livestock of that district, is disseminated to farming community to reduce the risk due to the vagaries of weather. An effort has been made to investigate the building of IT-based system to improve the efficiency of agromet bulletin. In this paper, we have explained the prototype of eAgromet, which is an IT-based agro-meteorological advisory system, to improve the process of agromet bulletin preparation. The concepts of weather deviation, similar agromet advice, and weather summary were conceived in eAgromet. The agromet experts have prepared the agromet bulletins using the prototype. The overall feeling by agromet experts is very positive. The system is very easy to use and can be deployed in any region. The agromet experts can prepare the agromet bulletins in the local language without any difficulty. The system is able to display the similar advice.

As a part of future work, we will make an effort to refine the prototype by taking into account all types of crops and livestock. By investigating the framework for crop-specific content development including the steps to deal with extreme weather events and contingency plans, we will also make an effort to minimize the human intervention for preparing agromet bulletin. Currently, IMD is planning to deliver agromet bulletins based on block-level weather forecast. We will make efforts to scale eAgromet for delivering block-level bulletins by minimizing human intervention and cost.

It is hoped that, for a given crop and region, after developing agromet bulletins for a few years, the effort to prepare new agromet bulletin could be reduced significantly due to efficient search. The agromet advice along with weather deviation and weather summary is a complex document which is prepared in local language. We are planning to investigate intelligent approaches to extract similar agroment advices for a given weather deviation and crop.

In the eAgromet prototype, we have defined the tags such that agromet expert could comprehend the weather trend and is able to prepare the agromet advice by analyzing how the weather deviation will influence the crop or live stock. Alternatively, the tags could be defined by considering each crop/livestock. For this, we will make an effort to build the system by considering cardinal weather values for all crops and livestock and define crop-specific tags. It will enable the building of intelligent agromet advisory system.

In addition to MRF, India Meteorological Department is also giving the following weather forecasts: now-casting, very short range forecast, short range forecast, and long range forecast. It is interesting to explore the building of eAgromet systems to deliver advisory information to farmers based on other types of weather forecasts.

Acknowledgements. The work is carried out as a part of the research project entitled "eAgromet: ICT-enabled Integrated Agro-Meteorological Advisory System" funded by India Meteorological Department/Ministry of Earth Sciences.

References

1. Farmers' Portal (January 2014), http://farmer.gov.in
2. eAgromet: An IT-based Agro-Mateorological Advisory System (January 2014), http://eagromt.in
3. Digital Green (January 2014), http://www.digitalgreen.org
4. Kisan Call Centers (January 2014), http://www.manage.gov.in/kcc.htm
5. eSagu: An IT-based Personalized Agro-advisory System (January 2014), http://www.esagu.in
6. Krishna Reddy, P., Ankaiah, R.: A Framework of information technology based agricultural information dissemination system to improve crop productivity. Current Science 88(12), 1905–1913 (2005)
7. Jones, J.W., Hansen, J.W., Royce, F.S., Messina, C.D.: Potential benefits of climate forecasting to agriculture. Agriculture, Ecosystems and Environment 82, 169–184 (2000)
8. Meinke, H., Hammer, G.L., Selvaraju, R.: Using Seasonal Climate Forecasts in Agriculture - The Australian Experience. 'Proof of Concept' or 'Taking the Next Step: Concept Adaptation?'. In: International Workshop on Climate Prediction and Agriculture (CLIMAG), Geneva, pp. 27–29 (1999)
9. Meinke, H., Hammer, G.: Experiences in Agricultural Applications of Climate Predictions: Australasia. In: Proceedings of the International Forum on Climate Prediction, Agriculture and Development. IRI-CW/00/1, pp. 52–58. International Research Institute for Climate Prediction, Palisades (2000)
10. Maini, P., Rathore, L.S.: Economic impact assessment of the Agrometeorological Advisory Service of India. Current Science 101(10), 1296–1310 (2011)

11. Ratnam, B.V., Krishna Reddy, P., Reddy, G.S.: eSagu: An IT based personalized agricultural extension system prototype - Analysis of 51 farmers case studies. International Journal of Education and Development using ICT (IJEDICT) 2(1), 79–94 (2006)

12. Krishna Reddy, P., Bhaskar Reddy, B., Kumaraswamy, M.: Village-Level Esagu: A Scalable And Location-Specific Agro-Advisory System. In: Proceedings of Third National Conference on Agro-Informatics and Precision Agriculture 2012 (AIAP2012), Hyderabad, India, pp. 47–52. INSAIT, Allied Publishers (2012)

13. Krishna Reddy, P., Bhaskar Reddy, B., Rama Rao, D.: A model of virtual crop labs as a cloud computing application for enhancing practical agricultural education. In: Srinivasa, S., Bhatnagar, V. (eds.) BDA 2012. LNCS, vol. 7678, pp. 62–76. Springer, Heidelberg (2012)

14. Reddy, D.R., Sreenivas, G., Mahadevappa, S.G., Rao, S.B.S.N., Varma, N.R.G.: Performance of CERES and WOFOST models in prediction of phenology and yield of rice in Telangana region of Andhra Pradesh. Journal of Agrometeorology (special issue - part I), 109–110 (2008)

15. Krishna Reddy, P., Bhaskar Reddy, B., Gowtham Srinivas, P., Kumaraswamy, M., Raji Reddy, D., Sreenivas, G., Mahadevaiah, M., Rathore, L.S., Singh, K.K., Chattopadhyay, N.: eAgromet: A Prototype of an IT-Based Agro-Meteorological Advisory System. In: The 8th Asian Federation for Information Technology in Agriculture (AFITA 2012), Taipei, Taiwan (2012)

16. Mahadevaiah, M., Raji Reddy, D., Sashikala, G., Sreenivas, G., Krishna Reddy, P., Bhaskar Reddy, B., Nagarani, K., Rathore, L.S., Singh, K.K., Chattopadhyay, N.: A Framework to Develop Content for Improving Agromet Advisories. In: Eigth Asian Conference for Information Technology in Agriculture (AFITA 2012), Taipei, Taiwan (2012)

17. Messina, C.D., Hansen, J.W., Hall, A.J.: Land allocation conditioned on El Nino-Southern Oscillation phases in the Pampas of Argentiana. Agricultural Systems 60, 197–212 (1999)

18. Ramakrishna, Y.S.: Current status of agrometeorological services in South Asia, with special emphasis on the Indo-Gangetic Plains, Working Paper No. 53, CGIAR Research Program on Climate Change, Agriculture and Food Security, CCAFS (2013)

19. Breuer, N.E., Cabrera, V.E., Ingram, K.T., Broad, K., Hildebrand, P.E.: AgClimate: a case study in participatory decision support system development. Climatic Change 87(3-4), 385–403 (2008)

20. Sivakumar, M.V., Hansen, J.: Climate prediction and agriculture: advances and challenges. Springer (2007)

MARST: Multi-Agent Recommender System for e-Tourism Using Reputation Based Collaborative Filtering

Punam Bedi, Sumit Kumar Agarwal, Vinita Jindal, and Richa

Department of Computer Science, University of Delhi
punambedi@ieee.org, sumitsagarwal@rediffmail.com,
vjindal@keshav.du.ac.in, richasingh.bv@gmail.com

Abstract. This paper presents a Multi-Agent Recommender system for e-Tourism (MARST) for recommending tourism services to the users. This system uses Reputation based Collaborative Filtering (RbCF) algorithm that augments reputation to existing Collaborative approach for generating relevant recommendations and to handle cold-start new user problem in tourism domain. The structure of a tourist product is more complex than a book or a movie and hence user profile modeling for these systems is much harder than most of other applications domains like books or movies. Moreover the frequency of activities and rating in tourism domain is also much smaller than in most of the other domains. This increases the complexity in designing and development of Recommender Systems in tourism domain. An attempt has been made in this paper to generate relevant services for a user in tourism domain using reputation based collaborative filtering. Most of the existing Recommender systems focus on one service at a time, whereas the proposed system incorporates three services (hotels, places to visit and restaurants) at a single place to ease the searching of information at one place only. The prototype of MARST has been designed and developed using various JAVA technologies and its performance was evaluated using precision, recall and F_1 metrics.

Keywords: Multi-Agent System, Recommender System, e-Tourism, Reputation.

1 Introduction

As information mounts, it leads to the problem of how to select desired information from the pool of available options. One possible solution to this information overload problem is recommender system, which employs information filtering techniques to assist users by giving personalized product or service recommendations. It offers recommendations of various items to users based on their preferences [22]. In this age of technology, recommender systems are usually found on websites with millions of visitors where users can receive recommendations for movies, websites, books, videos and music etc. [14]. Recommender systems are usually developed for e-Commerce applications. But at present, recommender system is gaining the popularity in the tourist domain where personalized services are offered to the travelers.

A. Madaan, S. Kikuchi, and S. Bhalla (Eds.): DNIS 2014, LNCS 8381, pp. 189–201, 2014.

Tourism is an activity that majorly depends on the personal interests and preferences of people. There are many websites to provide assistance in planning holidays with information on hotels, restaurants and places to visit. But these are not considered as successful applications as applications available for purchasing books or suggesting movies. The reason for this is maybe tourism is not a frequent activity and rating available here is much lesser than any other domain. Modeling user profiles is also a challenging task. Moreover Poor recommendations may result in valuable time being wasted, and therefore it is crucial for a recommender system for tourists to provide even the first recommendation given is of high quality. This problem is also handled in the proposed work. There are several ways of creating personalized recommendations for users. The existing recommender systems generally use collaborative, content-based, knowledge-based, hybrid approaches for generating recommendations regardless of reputation of items. In real scenario, reputation plays an important role in a decision taken by the user for accepting service recommendation. In the presented work, reputation of items (hotel, restaurants and places) is being incorporated within the recommendation process to generate relevant information for the user.

Majority of the recommender systems in tourism domain use collaborating filtering approach which is based on the idea that recommendations of items for a target user can be generated based on ratings given by the similar users. Recommender systems based on collaborative filtering approach face cold-start new user problem which occurs due to need of generation of recommendations for a user that has either not rated any item or rated very few items [1], [10]. In this case, the system is unable to compare target user with other users to find similar user for generating recommendations. In this paper, we propose Multi-Agent Recommender System for e-Tourism (MARST), which uses Reputation based Collaborating Filtering (RbCF) algorithm to reduce cold-start new user problem. The proposed RbCF algorithm augments the reputation of items within recommendation process to generate relevant services to the user.

The rest of the paper is organized as follows: Section 2 reviews the related work in the area. The proposed approach MARST with a detailed description of each step is presented in section 3. Experimental details and evaluations are shown in section 4 and finally section 5 concludes the paper.

2   Related Work

A lot of research has been done in tourism domain and recommender systems. As this domain need effective and accurate recommendation to attract the users. It would be of great help to all people to have the updated information about hotels, restaurants and places in a city such as cheap and best eating outlets or shelters or places which they can visit. The inevitable relationship between tourists, information and notes exist and it played a very important role in decision-making and became the foundation for the world economy [28]. Due to such importance of tourism industry, the recommendation systems applied in tourism have been a field of study.

The recommendation systems are an attempt to mathematically model and technically reproduce the process of recommendations in the real world [5], [21]. An intelligent recommendation system [2], [13], [18] is developed to provide a list of items that fulfill mostly all requirements in tourism domain. Many researchers introduced recommender system that deal with a case-based reasoning in order to help the tourist in defining a travel plan [24], [27]. The most promising recommendation systems in the tourism domain were knowledge-based and conversational approaches [25], [26].

Some other researchers proposed variants of the content-based filtering and collaborative filtering like knowledge-filtering, constraint-based and case-based approaches [15], [23], [29]. There also exist a recommendation system based on text mining techniques between a travel agent and a customer through a private Web chat [16]. A personalized tourist information provider is introduced as a combination of an event-based system and a location-based service applied to a mobile environment in [4], [11].

The current research in the area resulted in the improvement in the dependability of recommendations by certain semantic representation of social attributes of destinations [9]. There are some studies that focused on selecting the destination from a few exceptions [8], [20].

Some of the existing recommender system approaches offer support only for single aspects of a trip like restaurants [7] or attractions [12]. The proposed work combines three major services of tourism domain viz. hotels, restaurants and places in one application.

Although the considerable amount of work has been done on recommender systems in tourism domain, providing relevant service recommendation to the user is still a challenging task. We propose a system MARST, which augments reputation of items within recommendation process to generate relevant service recommendations using multi-agent approach. The system agents of MARST are working in a distributed and cooperative way, sharing and negotiating knowledge with the global objective of recommending the best prediction to the user. MARST provides the user with personalized recommendations and location-based information. The proposed approach focuses on recommending a list of the services that a tourist can get in a city like Delhi.

3   Proposed Multi-Agent Recommender System for e-Tourism

In this section, we present multi-agent recommender system for e-tourism (MARST) that automates the process of recommending restaurants, places to visit and hotels to the user. Section 3.1 presents the architecture of MARST system. The recommendation generation process of MARST is illustrated in section 3.2.

3.1   Architecture of MARST

The proposed system MARST is a multi-agent system, which contains a set of agents: specialized agents (Data Repository Agent (DRA), Items Reputation Agent (IRA) and

Similarity Agent (SA)) and the system agents (User Agent (UA), Hotel Recommendation Agent (HRA), Restaurant Recommendation Agent (RRA) and Place Recommendation Agent (PRA)). All specialized agents execute in the background to periodically compute and store the required information within data repository for future recommendation generation. The system agents are responsible for computing recommendations by utilizing the information stored within data repository on user's request. The specialized agent DRA deals with extracting and aggregating information from the web services. The other two specialized agents SA and IRA compute similarity between users and reputation of all items respectively. The system agents interact with each other to generate recommendations using proposed reputation based collaborative filtering algorithm (RbCF).

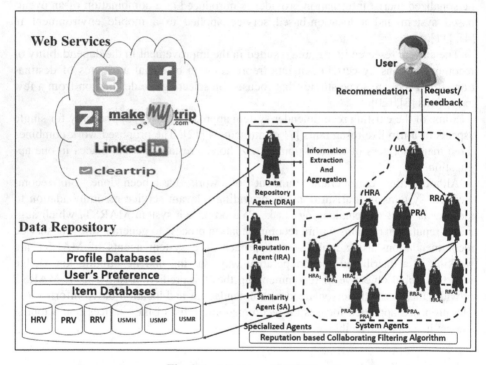

Fig. 1. Architecture of MARST

The system agent UA represents the user who wants a recommendation from the system. It is created, whenever a new user registers with MARST. The basic functionality of UA is to infer and keep information about the registered user. UA receives a query request of service recommendation from the user using web based GUI and then passes it to either HRA, or PRA, or RRA or the combination of two or all at the same time for generating recommendations. After receiving the recommendation list from the requested recommendation agent, the UA finally displays it to the user. The recommendation agents HRA, RRA and PRA are responsible for computing service recommendations (hotels, restaurants and places to visit respectively) for the UA on its request. The recommendation agents further call a subagent of same type that interacts with data repository to compute recommendations. The subagent in the system is automatically created and terminated for a particular service recommendation request. Figure 1 illustrates the architecture of MARST system.

3.2 Recommendation Generation Process of MARST

In MARST, recommendations are generated using reputation based collaborative filtering (RbCF) algorithm. This algorithm works in two processes: offline process and online process. In offline process, the specialized agents periodically compute and store the items reputation vectors (HRV, RRV and PRV) and user similarity matrices (USMH, USMR and USMP) within data repository for future recommendation generation. In online process, the recommendations are generated by the system agents for the target user.

RbCF algorithm is briefly outlined as follows:

Offline Process

1. Input data normalization.
2. Create items reputation vectors (HRV, RRV and PRV), user similarity matrices (USMH, USMR and USMP) and store this information within data repository.

Online Process

1. Select similar users and aggregate their recommendation lists.
2. If an aggregated list obtained from step1 contains at least N items to be recommended then go to step5 else go to step3.
3. Filter items from appropriate item reputation vector based on the distance specified by the user in his query request.
4. Aggregate both lists obtained from step1 and step3.
5. Display top N items of the aggregated list to the target user.

Steps listed in the offline and online processes are explained below in detail.

Offline Process:
Step1) Input Data Normalization:
User's preferences are stored within data repository in the form of user-item rating matrices for each recommendation service (hotel, restaurant and place) separately. In these matrices users represent the rows, items represent the columns and the value within i^{th} row and j^{th} column represents the rating of ith user about the jth item. These user-item matrices consist of ratings in the scale 1 to 5. The system normalizes these ratings in the range 0 to 1 as follows:

$$r_{ij} = \frac{r_{ij}}{\sum_{j=1}^{n} r_{ij}} \qquad (1)$$

where
r_{ij} represents the rating of ith user for j^{th} item within user-item rating matrix
n denotes the total number of items within user-item rating matrix

Step2) Create items reputation vectors (HRV, RRV and PRV), user similarity matrices (USMH, USMR and USMP) and store this information within data repository:

(i) The reputation of each item is computed by the specialized agent IRA to form Hotel Reputation Vector (HRV), Restaurant Reputation Vector (RRV) and Place Reputation Vector (PRV) from the corresponding user-item rating matrices. The reputation of jth item ROI_j [3] in these vectors is computed as follows:

$$ROI_j = \left(\frac{3 \times avg_j \times \binom{n_j}{N} \times \left(\frac{1}{SD_j}\right)}{avg_j \times \binom{n_j}{N} + \binom{n_j}{N} \times \left(\frac{1}{SD_j}\right) + avg_j \times \left(\frac{1}{SD_j}\right)} \right) \qquad (2)$$

where
avg_j represents the average rating of jth item
n_j represents number of users who rated j^{th} item in the user-item rating matrix
N denotes total number of users in user-item rating matrix
SD_j denotes standard deviation of the ratings given by individual users for the jth item.

(ii) The similarity between users are computed by specialized agent SA to form user similarity matrix for hotels (USMH), user similarity matrix for restaurants (USMR) and user similarity matrix for places (USMP) from the corresponding user-item matrices. The Pearson correlation coefficient metric [6] has been used by SA to compute these matrices.

Online Process:

The online process starts, when the system agent UA receives a query request of service recommendation from the user. Thereafter, UA passes that query to either HRA, or PRA, or RRA or the combination of two or all at the same time depending on the type of service or services requested by the user. The requested recommendation agent further calls a subagent of same type that interacts with data repository to generate recommendations. The subagent gives the generated recommendations to the requested recommendation agent, which further passes it to UA. Finally, UA displays the received recommendations to the user. The steps used by subagent to generate recommendations are described in detail below:

Step1) Select similar users and aggregate their recommendation lists:

The subagent selects the similar users from USMH or USMR or USMP based on the service requested by the user. Then it retrieves the rated items of selected similar users from the corresponding user-item rating matrix. Thereafter, subagent prepares an aggregated list from the retrieved item lists of similar users as below:

(i)   Identify distinct items from the retrieved item lists of similar users.
(ii)   Compute degree of importance (DOI$_j$) for each distinct item I$_j$ as below:

$$DOI_j = ROI_j \times DD(u, I_j) \tag{3}$$

where
        ROI$_j$ denotes the reputation of jth item
        DD(u, I$_j$) denotes the distance decay function [3] for the item I$_j$ and the target user u. The objective of this function is to minimize the degree of importance of an item, when the current location of target user increases from that item.

Arrange the items with nonzero DOI in descending order of their DOI.

Step2) If an aggregated list obtained from step1 contains at least N items to be recommended then go to step5 else go to step3:
Subagent determines the number of items available within aggregated list. If it is greater or equal to N items to be recommended, then subagent recommends top N items of the aggregated list to the requested recommendation agent; otherwise it executes next step.

Step3) Filter items from appropriate item reputation vector based on the distance specified by the user in his query request:
The subagent filters the items from HRV or RRV or PRV depending on the type of service requested by the user. It filters and aggregates the items using following steps:

Filters those items, whose reputation value is greater than some threshold.
Computes DOI for each filtered items using equation (3).
Arranges the filtered items in descending order of their DOI.

Step4) Aggregate both lists obtained from step1 and step3:
The item lists obtained from step1 (list1) and step3 (list2) are aggregated as follows:

Insert list1 at the top of aggregated list.
Remove common items of both lists from list2 then remaining items of list2 are appended within aggregated list.

Step5) Recommend top N items of the aggregated list to the requested recommendation agent:
The subagent recommends top N items of the aggregated list to the requested recommendation agent, which further passes it to UA. Finally, UA displays the received recommendations to the user.

4   Experimental Details

A prototype of the MARST was designed and developed using JSP (Java Server Pages) for creating user interface, JADE (Java Agent Development Environment) for constructing multi-agent environment and MySql 5.0.24 for building data repository. Figure 2 depicts the user's request for service recommendation from the system. The computations during recommendation generation are shown in figure 3. The recommendation generated by the system in response to the user's request is illustrated in figure 4.

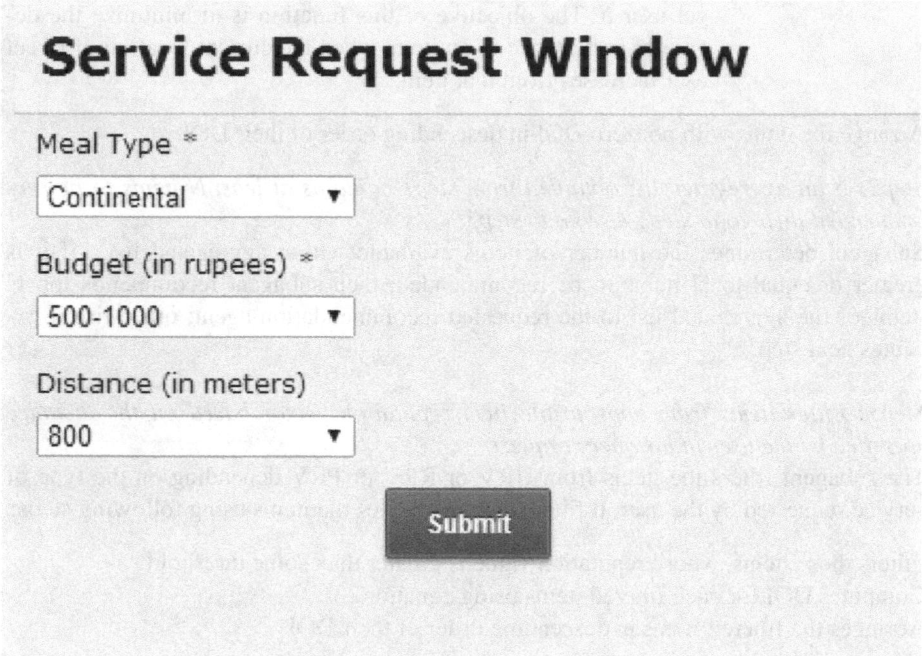

Service Request Window

Meal Type *

| Continental ▼ |

Budget (in rupees) *

| 500-1000 ▼ |

Distance (in meters)

| 800 ▼ |

Submit

Fig. 2. Snapshot represents the user's request for service recommendation

The performance of implemented prototype was evaluated using standard metrics precision, recall and F1 measure, which focus on quality and accuracy of recommendations. Precision is analogous to positive predictive value. Recall score of 1.0 indicates that all relevant recommendations were retrieved. One of the ways to evaluate precision and recall is to predict the top N items for recommendation. Recall and precision for top N recommendations can be defined as follows:

Precision is obtained as a ratio between the number of items both relevant, retrieved in the top N recommendation list and the number of items retrieved by the system. It gives the average quality of an individual recommendation. Recall is obtained as a ratio between the number of items both relevant, retrieved in the top N

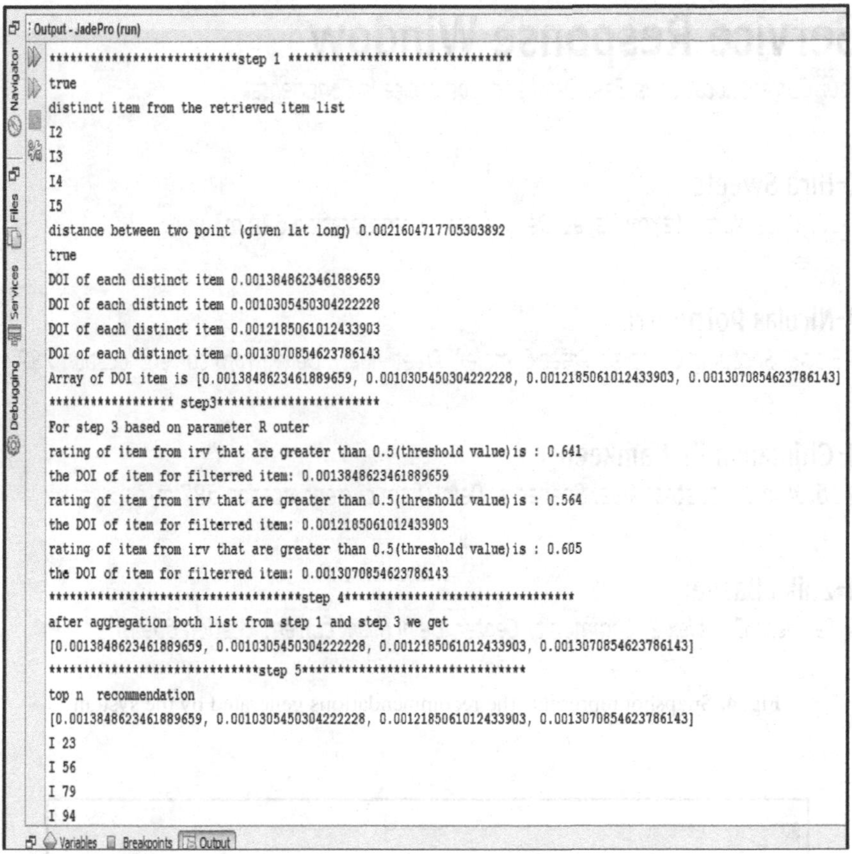

```
Output - JadePro (run)
******************************step 1 ********************************
true
distinct item from the retrieved item list
I2
I3
I4
I5
distance between two point (given lat long) 0.0021604717705303892
true
DOI of each distinct item 0.0013848623461889659
DOI of each distinct item 0.0010305450304222228
DOI of each distinct item 0.0012185061012433903
DOI of each distinct item 0.0013070854623786143
Array of DOI item is [0.0013848623461889659, 0.0010305450304222228, 0.0012185061012433903, 0.0013070854623786143]
****************** step3*************************
For step 3 based on parameter R outer
rating of item from irv that are greater than 0.5(threshold value)is : 0.641
the DOI of item for filterred item: 0.0013848623461889659
rating of item from irv that are greater than 0.5(threshold value)is : 0.564
the DOI of item for filterred item: 0.0012185061012433903
rating of item from irv that are greater than 0.5(threshold value)is : 0.605
the DOI of item for filterred item: 0.0013070854623786143
***********************************step 4*************************************
after aggregation both list from step 1 and step 3 we get
[0.0013848623461889659, 0.0010305450304222228, 0.0012185061012433903, 0.0013070854623786143]
*******************************step 5***********************************
top n  recommendation
[0.0013848623461889659, 0.0010305450304222228, 0.0012185061012433903, 0.0013070854623786143]
I 23
I 56
I 79
I 94
```

Fig. 3. Shows the computations during recommendation generation

recommendation list and the number of items actually relevant. These two measures are clearly conflicting in nature. If the number of recommendations (top N) produced increases, then the value of recall increases, while at the same time precision is decreased. But since both recall and precision are important in evaluating the performance of a system that generates top N recommendations, they can be combined with equal weights to get a single metric, the F1 metric. Higher values of F1 indicate a more balanced combination between recall and precision.

The performance of RbCF algorithm was also compared against conventional CF approach on cold start users (who rated 0 to 4 items) using Mean Average Precision (MAP) and User Coverage metrics. The MAP for N users is defined as arithmetic mean of the average precision of each user [19]. The user coverage [17] defined as the portion of users for which the system is able to predict at least one item. The results are shown in Table 1.

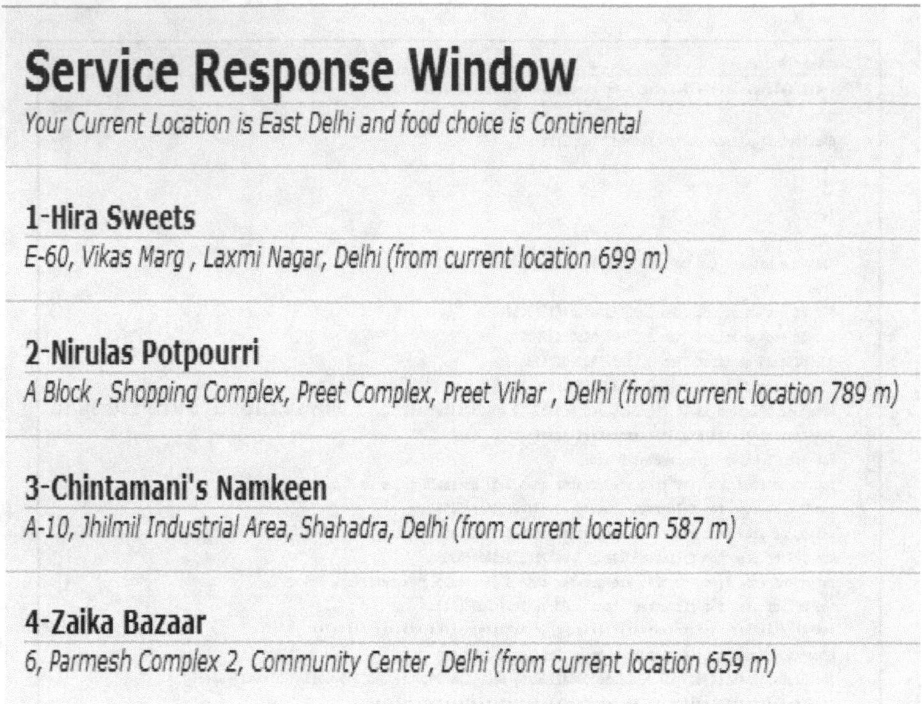

Service Response Window
Your Current Location is East Delhi and food choice is Continental
1-Hira Sweets
E-60, Vikas Marg , Laxmi Nagar, Delhi (from current location 699 m)
2-Nirulas Potpourri
A Block , Shopping Complex, Preet Complex, Preet Vihar , Delhi (from current location 789 m)
3-Chintamani's Namkeen
A-10, Jhilmil Industrial Area, Shahadra, Delhi (from current location 587 m)
4-Zaika Bazaar
6, Parmesh Complex 2, Community Center, Delhi (from current location 659 m)

Fig. 4. Snapshot represents the recommendations generated by the system

90
80
70
60
50
40
30
20
10
0
Top N = 3
Top N = 5
Top N = 10
Precision
Recall
F1

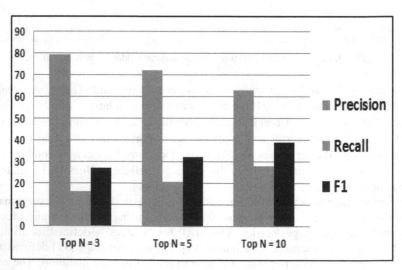

Fig. 5. Performance of MARST using Precision, Recall and F1 metric at top N= 3, 5, 10 (Here Precision, Recall and F1 metrics are in percentage)

Table 1. Results of experiment on cold start users who rated 0 to 4 items. Rows represent the evaluation measures, we collected for the RbCF and conventional CF approaches. Columns represent different views of the data (e.g., in the last column, we present evaluation measures computed only on users who have rated exactly 4 items).

Ratings		0	1	2	3	4
# Cold Start		4	12	21	32	37
MAP	CF	0	0	0.081	0.292	0.37
	RbCF	0.37	0.43	0.48	0.521	0.53
User	CF	0%	0%	10.13%	16.4%	26.04%
Coverage	RbCF	100%	100%	100%	100%	100%

5   Conclusion

A Multi-Agent Recommender System for e-Tourism (MARST) is presented in this paper. MARST augmented the reputation of items within its recommendation process using proposed RbCF algorithm to generate relevant services recommendations to the user. The recommendations for three major services of tourism domain viz. hotels, restaurants and places-to-visit are handled in a single application. The cold start new user problem is also handled within RbCF. The system agents of MARST work in a distributed and cooperative way, sharing and negotiating knowledge with the global objective of recommending the best prediction to the user. MARST provides the user with personalized recommendations and location-based information. MARST has been implemented using various java technologies. The results obtained from the experiments were compared with results of conventional collaborative filtering approach and it was found that RbCF performs better than conventional collaborative filtering approach.

References

1. Ahn, H.J.: A new similarity measure for collaborative filtering to alleviate the new user cold-starting problem. Information Sciences, 37–51 (2008)
2. Bansal, M., Mirzadeh, N., Ricci, F.: Supporting User Query Relaxation in a Recommender System. In: 5th International Conference on E-Commerce and Web Technologies (EC-Web), Zaragoza, Spain, pp. 31–40 (2004)
3. Bedi, P., Agarwal, S.: SRPRS: Situation-Aware Reputation Based Proactive Recommender System. Journal of Information Assurance and Security 8, 220–229 (2013)
4. Bedi, P., Agarwal, S.K.: Situation Aware Proactive Recommender System. In: 12th International Conference on Hybrid Intelligent Systems, pp. 85–89. IEEE Xplore, Pune (2012)

5. Bedi, P., Agarwal, S.K.: Aspect-oriented Trust Based Mobile Recommender Systems. International Journal of Computer Information Systems and Industrial Management Applications, 354–364 (2013)
6. Benesty, J., Huan, Y., Chen, J.: Pearson Correlation Coefficient. In: Benesty, J., Huan, Y., Chen, J. (eds.) Noise Reduction in Speech Processing. Springer Topics in Signal Processing, pp. 1–4. Springer, Heidelberg (2009)
7. Burke, R.: Knowledge-based recommender systems. Encyclopedia of Library and Information Systems 69 (2000)
8. Charou, E., Kabassi, K., Martinis, A., Stefouli, M.: Integrating Multimedia GIS Technologies in a Recommendation System for Geo-tourism. In: Tsihrintzis, G.A., Jain, L.C. (eds.) Multimedia Services in Intelligent Environments. Smart Innovation, Systems and Technologies, vol. 3, pp. 63–74. Springer, Heidelberg (2010)
9. Daramola, O.J., Adigun, M.O., Ayo, C.K., Olugbara: Improving the Dependability of Destination Recommendations Using Information on Social Aspects. Tourismos: An International Multidisciplinary Journal of Tourism 5(1), 13–34 (2010)
10. Freyne, J., Berkovsky, S., Smith, G.: Rating Bias and Preference Acquisition. ACM Transactions on Interactive Intelligent Systems (2013)
11. Hinze, A., Voisard, A., Buchanan, G.: TIP: Personalizing Information Delivery in a Tourist Information System. Journal of Information Technology & Tourism 11(3), 247–264 (2009)
12. Huang, Y., Bian, L.: A Bayesian network and analytic hierarchy process based personalized recommendations for tourist attractions over the internet. Expert Systems with Applications, 933–943 (2009)
13. Jannach, D.: Finding Preferred Query Relaxations in Content-based Recommenders. In: IEEE Intelligent Systems Conference, pp. 355–360. IEEE, Westminster (2006)
14. Jannach, D., Zanker, M., Felfernig, A., Friedrich, G.: Recommender Systems - An Introduction. Cambridge University Press, NY (2011)
15. Kazienko, P., Kolodziejski, P.: Personalized Integration Recommendation Methods for E-commerce. International Journal of Computer & Applications 3(3), 12–26 (2006)
16. Loh, S., Lorenzi, F., Saldaña, R., Licthnow, D.: A Tourism Recommender System Based on Collaboration and Text Analysis. Information Technology & Tourism 6, 157–165 (2004)
17. Massa, P., Avesani, P.: Trust-aware recommender Systems. In: Proc. of ACM Recommender Systems, pp. 17–24 (2007)
18. McSherry, D.: Retrieval Failure and Recovery in Recommender Systems. Artificial Intelligence Review (24), 319–338 (2005)
19. Minkov, E., Charrow, B., Ledlie, J., Teller, S., Jaakkola, T.: Collaborative Future Event Recommendation. In: CIKM 2010, pp. 1–9. ACM (2010)
20. Niaraki, A.S., Kim, K.: Ontology Based Personalized Route Planning System Using a Multi-criteria Decision Making Approach. Expert Systems with Applications 36, 2250–2259 (2009)
21. Petrevska, B., Koceski, S.: Tourism Recommendation System: Empirical Investigation. Journal of Tourism (14), 11–18 (2012)
22. Ricci, F., Rokach, L., Shapira, B., Kantor, P.B.: Recommender Systems Handbook. Springer, New York (2011)
23. Ricci, F., Del, F.M.: Supporting Travel Decision Making through Personalized Recommendation. In: Karat, C.-M., Blom, J., Karat, J. (eds.) Designing Personalized User Experiences for e-Commerce, pp. 221–251. Kluwer Academic Publishers (2004)

24. Ricci, F., Werthner, H.: Case Base Querying for Travel Planning recommendation. Information Technology & Tourism 4(3/4), 215–226 (2002)
25. Ricci, F., Arslan, B., Mirzadeh, N., Venturini, A.: ITR: A Case-based Travel Advisory System. In: 6th European Conference on Advances in Case-Based Reasoning, Europe, pp. 613–627 (2002)
26. Thompson, C., Göker, M., Langley, P.: A personalized system for conversational recommendations. Journal of Artificial Intelligence Research 21, 393–428 (2004)
27. Wallace, M., Maglogiannis, I., Karpouzis, K., Korm: Intelligent One-stop-shop Travel Recommendations Using an Adaptive Neural Network and Clustering of History. Information Technology & Tourism 6, 181–193 (2003)
28. Wang, J.: Improving decision - making practices through information filtering. International Journal of Information and Decision Sciences 1(1), 1–4 (2008)
29. Zanker, M., Fuchs, M., Höpken, W., Tuta, M.M.: Evaluating Recommender Systems in Tourism - A Case Study from Austria. In: O'Connor, P., et al. (eds.) Proceedings ENTER 2008, Information and Communication Technologies in Tourism, pp. 24–34. Springer (2008)

A Model of Privacy and Security for Electronic Health Records

Pulkit Mehndiratta, Shelly Sachdeva, and Sudhanshu Kulshrestha

Jaypee Institute of Information Technology
{pulkit.mehndiratta,shelly.sachdeva,sudhanshu.kulshrestha}@jiit.ac.in

Abstract. Information and communication technology has created excellent development in over the past few years in the field of medicine and healthcare. Healthcare is constantly undergoing changes, with new medical technologies, business models and research findings. The requirements for security and privacy are also very critical and very difficult to satisfy in case of Electronic Health Records (EHRs) data especially as compared to any other data. This is due to the conflicting needs of clinicians (who demand open and easy access to databases) and the patients (who prefer closed and private access to information stored in databases). The potential and capabilities of IT and its influence on the Indian healthcare is of utmost importance. Thus, this study examines the current status of security and privacy of various healthcare services/solutions implemented for electronic health records in India. This topic has not been sufficiently addressed by the existing healthcare solutions based on standards. The authors aim to bridge this gap by proposing a model to protect the security and privacy for Standardized Electronic Health Records EHRs database systems. A simulative analysis for the implementation of the proposed model has been presented. This will help in large scale deployment of secured Electronic Health Record systems that will benefit hospitals and their users.

Keywords: Security and Privacy, Electronic Health Record, Developing Country, India.

1 Introduction

1.1 Electronic Health Records

The medical domain is vast and complex. Electronic Health Records (EHRs) are the paperless solution to a disconnected healthcare world that runs on a chain of paper files. They provides new opportunities, improves productivity, reduces the administrative burdens, reduce cost and medical errors. These become cavillous in the case of an emergency where the patient may be unable to communicate this information. These provide doctors with more timely access to potentially life-saving information at the point of care while diminishing the paper trail.

In general, an EHRs database includes lifelong history of medical documents for any person which includes clinical statements such as observations, laboratory

A. Madaan, S. Kikuchi, and S. Bhalla (Eds.): DNIS 2014, LNCS 8381, pp. 202–213, 2014.

tests, diagnostic imaging reports, treatments, therapies, drugs administered, and allergies. Thus, over a period of time the size of the database becomes very large and very fast leading to various management and security issues.

"As more of our medical records are stored electronically, the threats to our security and privacy increase"[1]. Electronic health records form an integral part of the healthcare system and it is imparitive that EHRs are safe because there is evidence that breaches in security have an impact on patients health care. In a survey conducted in 2006 [11], 62% of the public said "The use of electronic medical records makes it more difficult to ensure patients' privacy". However similar proportions recognized the potential for EHRs in cost and error reductions and increased patient safety. Thus, unless privacy and security problems are resolved, EHRs will not be widely adopted.

1.2 Privacy and Security

The definition of privacy emphasizes the control over the Personally Identifiable Information that should always rest with the data subject. Taking control over this information/data from the subjects takes away his/her privacy. Whereas, security is defined as the extent to which this personally identifiable information can be stored and shared in such a manner that access to the information is limited to authorized parties.

On one hand, best solution to protect the privacy is that data subject should volunteer their own information as they may want to delegate only some (or all) the controls to others. On the other hand, security to the EHRs systems can be provided by the means of physical security of the system, using access control mechanism, or by the use of firewalls and encyrption techniques.

Four major threats (to privacy of data) identified by US National Research Council [23] on medical privacy refer primarily to insider attacks. Many systems like Microsoft Health Vault and Google Health comply with data protection acts by letting the patients decide on the usage and disclosure of their data. But these fail in satisfying essential requirements to privacy and security.

In United States, much of the sensitive data such as insurance information, sensitive patient communications, and personally identifiable information are protected under the U.S. law (the Health Information Technology for Economic and Clinical Health (HITECH) Act). The legislation that regulates release of health-care information is Health Insurance Portability and Accountability Act (HIPAA)[3]. Where on one hand EHRs have to follows strict guidelines of the standards they are based upon, there are legal aspects also when it comes to personal identifiable information. According to Spanish law [12, 13], an article states that *"no personal data will be stored in files that do not meet the requirements of integrity and security"*.

Electronic Health Record data become critical in the case of an emergency where the patient may be unable to communicate this information. These provide doctors with more timely access to potentially life-saving information at the point of care while diminishing the paper trail.

In developing countries like India, the conventional system of medication is still restricted to paper and pen. With a population of over 1 billion, EHRs databases can be a boon to the existing system. EHRs represent lifelong documentation of medical history for any patient and the size of the database increases exponentially. So, an efficient protocol and model is required similar to what is proposed in [9,10]. Thus, it is of utmost importance to provide doctors and patients with modern facilities like computer and mobile based medical solution. This will ease the work of practitioners and make it more effective and productive. At the same time security and privacy of the data has to be maintained in the system. Few of the breaches that occurred in past six to eight months around the globe [4] are due to lack of security and privacy measures and it has affected the lives of patients. ISO/TS 18308 standard gives the definitions of security and privacy issue for EHRs [2].

This paper contributes to the current status of EHRs in India and what are the various security and privacy issues. Section 2 throws light on whether various EHRs implemented in India follow any standards, along with what are the guidelines to be taken into consideration while creating any EHRs solution. The security is required for all the components/layers encountered between the front-end and back-end. Section 3 details the privacy and security measure required before deploying any EHRs database system. A simulation of various techniques has also been discussed in this section providing detailed interaction between various layers on EHRs database systems. Section 4 present the discussion. Section 5 illustrate the conclusions.

2   Privacy and Security Concerns in Healthcare

The main difference for healthcare data as compared with any other industry is mainly the sensitivity of healthcare data. Patients want their private data to be not disclosed without their consent. Clinicians and researchers want access to patient data in order to come up with concrete solutions in their work. Information is therefore required to be provided in such a way that the personal identity information cannot be disclosed.

Some of the patients records were left exposed to public at University of Michigan Medical center on the internet because the center thought that they were on a server protected by a password [25]. Various other breaches have been reported in [4]. Thus, unless privacy and security problems are resolved, EHRs will not be widely adopted.

There are several standards for EHRs interoperability, such as the Health Level 7 (HL7) Clinical Document Architecture (CDA), CEN EN 13606 EHRcom, openEHR[15], Digital Imaging and Communications in Medicine Structured Reporting (DICOM SR), Web Access to DICOM Persistent Objects (ISO WADO), integrating the Healthcare Enterprise (IHE), Retrieve Information for Display (RID) and IHE Cross-Enterprise Document Sharing (XDS) [14].

ISO/TS 18308 standard gives the definitions of security and privacy issue for EHRs. The Working Group 4 of International Medical Informatics Association (IMIA) was set up to investigate the issues of data protection and security within the healthcare environment. Its work to date has mainly concentrated on security in EHRs networked systems and common security solutions for communicating patient data. The European AIM/SEISMED (Advanced Informatics in Medicine/Secure Environment for Information Systems in MEDicine) project is initiated to address a wide spectrum of security issues within healthcare and provides practical guidelines for secure healthcare establishment. The general technical standard on information security is under study by ISO/SC 27, while the medical-specialized technical standard and the guidelines for using such a standard are under development by ISO/TC 215 WG4, CEN/TC 251 WG3, and HL7 WG13. Figure 1 presents the security features of some existing standards in EHRs. It is prepared by studying enhancement (including two standards) of security features [14].

STANDARDS → SECURITY FEATURES ↓	CEN 13606	openEHR	DICOM (WADO)	DICOM (SR)	IHE (RID)	IHE(XDS)
Transport layer encryption support	YES	YES	YES	YES	YES	YES
Protocol allows to transmit user	YES	YES	YES	YES	YES	YES
Protocol enforces access rules	YES	YES	YES	NO	NO	YES

Fig. 1. Security Features of Standardized EHRs

Although, many architectures have been proposed in order to provide better security and privacy to the user but still almost all the systems are lacking in providing the same. The disparity between patients needs and desires for security & privacy and what is provided by some of the electronic health record systems, is illustrated by the results of a study commissioned by HSS [21], which found that the privacy policies of Personal Health Record (PHR) vendors, a type of health controlled by the patients, "lacked the standard components of privacy notices". Thus, one has to come with a system which is patient oriented, so that patient can use and interact with the system in easier manner and at the same time provides security and privacy [22].

2.1   Case Study

Healthcare is moving to the world of computing and the use of new technologies. The ability to quickly and pro-actively react to changes in the healthcare industry will likely become a necessity to stay ahead of the increasing demands of reducing costs, compliance mandates, security concerns and improving the quality of patient care. A standardized format and content of a patient's clinical record helps promote the integration and continuity of care among the various providers of care to the patient. If the data perceived conforms to healthcare standard(openEHR/ HL7), a vendor may conclude that data credibility is high. Keeping this in mind, the research's motive is to explore the privacy and security of standard based EHRs application for its use in India.

EHRs Services in India: In India, very few agencies are working towards area of Health Informatics and Electronic Health Records (EHRs). C-DAC (Center for Development of Advanced Computing) has developed various solutions such as *E-Sushrut [5], DIGHT [6], Mercury, E-Sanjeevni, Tejhas, Ayusoft* etc. Most of these solutions are indigenously developed and managed by C-DAC only. We have done an extensive study of the architecture of all the products and solutions developed and tried to evaluate the security and privacy component in it.

Among these, *E-Sushrut* [5] is the most comprehensive and widely deployed Health Information System. This system incorporates an integrated computerized clinical information system for improved hospital administration and patient health care. HISP India [18] is a training, analysis and evaluation organization creating HIS (Health Information Systems) for South East Asia and is headed by University of Oslo, Norway. HISP has developed few products which cater the EHRs solution requirement. For example, its flagship product DHIS2 which has been recently deployed in a block of Punjab in India. The real time version streamlines the flow of patients and simultaneously empowers workflow to perform to their peak ability, but the security and privacy of the patients data is only limited to the user-level access control mechanism. No attention has been paid to the data encryption and anonymity which could lead to inferences from the available information. The system also lacks in various measures to protect it from network attacks. Thus, very critical and highly confidential information can easily be compromised due to lack of proper measures.

Project DIGHT (Distributed Infrastructure for Global EHRs Technology) [6] proposed to have a separate module for security and privacy which will provide secure storage and access of EHRs, along with privacy to the user. But, till date no such module has been developed/implemented for India to suffice the purpose.

EHRs Standards in India: A report on EHRs standards for India approved by Ministry of Health and Family Welfare (MHFW) has been released in August 2013 [19]. It stated the following factors as shown in (figure 2) of data privacy and security to be considered while creating EHRs solution.

1. **Access Control:** A mechanism for uniquely identifying a user and enabling the user with control for authorized access of EHRs. Although, an exception could always be there for critical circumstances.

Fig. 2. Privacy and Security Functionalities for Creating EHRs solutions

2. **Access Privilege:** The access to an EHRs of a user should be dependent upon the varying privilege of the care provider(s), as specified by the EHRs administrator.
3. **Auto Log-off:** This is a feature which encapsulates the session management as well as state management. A session time-out should be defined and the states of the record thereof should be saved as intact.
4. **Audit-log:** A log should be maintained which would store each and every action performed over any record. It should include timestamp, user id & the action performed.
5. **Integrity:** The standards should not be compromised in the data transit and also, the deletion of any record as well as audit-log should be prohibited.
6. **Authentication:** User authentication mechanism should be deployed to check the authenticity of the user locally or on a network.
7. **Encryption:** A suitable mechanism should be used for same with substantial key length. Also, in cases of alteration and intrusion, the verification technique should be there e.g. hashing. Also, the data transaction should contain the substantial id information that the receiver can perform security audit trails along with access control decisions.

All the product designed and developed by C-DAC are lacking in security and privacy component. Thus, we have proposed a model to provide security and privacy to the end-user.

3   Proposed Model

There is no present case for India to maintain privacy and security of standardized EHRs database systems. However, MHFW has presented set of factors (please see section 2) to be included as a requirement for secure EHRs systems in the future. Considering various privacy and security concerns and their manifestations, the authors propose a model for EHRs database systems as shown in figure 3. We have tried to included the security at each reference layer of the standardized EHRs. Also, a simulation of various techniques that can be applied to the each layer has been discussed in the sections below.

3.1   Proposal for Secure Model

Our proposal as shown in figure 3,aims to provides security and privacy to the user (data subjects). This shows the function wise reference layer model of the EHRs system. The goal is, how we can include security and privacy techniques on each layer of this reference model of electronic health record database systems to give maximum security as well as state of the art privacy to the data subjects.

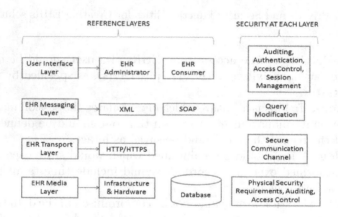

Fig. 3. Reference model for the Standardized Electronic Health Records Database systems with privacy and security measures at each layer

1. **The User Interface Layer:** It is the top most layer where the end user (doctors, nurses and pateints) is interacting with the system.In healthcare domain, the users vary in their background skills and have variable needs. Moreover, they interact with systems in various contexts. Techniques like access control and audit logging can be very helpful to provide adequate security to this layer. For example, if a patient is suffering from fever, or any minor disease, he may share his information, But, a patient will hesitate if he/she is suffering from any sexual or acute disease such as, tuberculosis. The consent of the patient should be taken and they should be familiarize with all the implications once the data is shared. Authentication mechanism also plays an important role when a person has to interact with the system. This is because unwanted entities can be stopped for accessing the system.

2. **The Messaging Layer:** This layer consists of the various messaging formats of the EHRs system which are mainly in XML (Extensible Markup

Language) or use SOAP (Simple Object Access Protocol) for exchanging the structured information through a web service. We need to specify various rules and regulations for the query evaluation purposes. in order to provide security to this later Iif any query is violating the rules or the guidelines then the modification technique should be used to protect the unwanted sharing of data. For example, controlled query evaluation enforces security policies for confidentiality in information systems [27, 28]. Also, the system must provide means where user queries can be modified. This is important in healthcare environment when the user wants to query data from another hospital.

3. **The Transport Layer:** This is most important layer for any EHRs service as all the request are forwarded via this layer itself. Using a secure communication channel i.e. HTTPS can make data-flow more secure as this protocol encrypts the data to be transferred over the web. Further, this protects the service from various attacks like packet sniffing.

4. **The Media Layer:** EHRs databases are complex, vast and temporal in nature. Media layer is holding the most important asset i.e., the actual database of the system. It also comprises of all the hardware and the infrastructure too. Privacy can be increased by storing data as anonymous as possible. Database separation can be done based upon the demographic information and the healthcare data. Access Control and audit logging play a critical role in securing the data from insider attacks. Moreover, the physical security of the database is also necessary.

Above mentioned reference model with security features can be applied to any standardized EHRs solution.

3.2   Simulation of Proposed Model Using Various Techniques

In the last section, we have discussed various security and privacy measures that can be included at each layer of any standardized electronic health record system thus, to make it more secure and usable for end user. This section presents simulation of privacy and security at various layers proposed in reference model through various techniques. Figure 4 shows how the various layers will interact with each other internally once the system is deployed for use. The tag cloud consists of various techniques (numbered from 1 t0 8 as shown in figure 4). It throws light on how the various techniques that are part of the tag cloud, are being used to suffice the goal of providing security and privacy to the EHRs systems. The figure detailswhich technique should be used in which layer for the smooth handling and storage of sensitive healthcare data. It also shows the interaction between different layers through secure communication channel, sending query and its retrieval with secure message being exchanged.

1. **The Demographic and Electronic Health Record Separation:** The database can be further separated into set of two databases. One database will store healthcare information and the other database will be having the

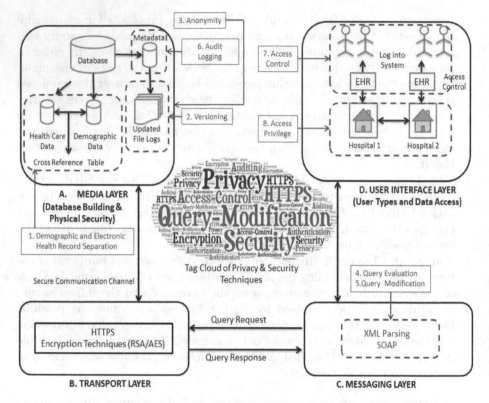

Fig. 4. Simulation for Proposed Model using Various Techniques

demographic details of the patients. A cross-reference table will be maintained to match the record from both the databases.

2. **Versioning:** Along with demographic separation, a meta data log should be maintained that contains the time stamp information when any changes are made to any of the records. At the same time, the new record should be declared as active while the previous record becomes inactive.

3. **Anonymity:** The use of anonymity techniques like k-anonymity and L-diversity will make databases more generalized. Thus, in case of crisis the loss of sensitive information may be minimized. Since, the L-diversity prevents the homogeneity attack and the background knowledge attack posed by k-anonymity, hence choosing L-diversity for sensitive health data would be better option for standardized EHRs systems.

4. **Query Evaluation:** User run queries should be evaluated based on the privacy policy and guidelines to make sure they do not violate it. In case of violation, appropriate measures like evaluation or modification of the query should be done. For example, controlled query evaluation enforces security policies for confidentiality in information systems [17],

5. **Query Modification:** The system must provide means where user queries can be modified [16]. This is important in healthcare environment where

querying medical information may vary depending on the role of end-user and the way information is protected. The e-Diamond project [24] overcomes the query modification issues (multiple views and removing order, primary keys and joins) for medical databases. Measures (taken in discussed method) will reject the query if it does not fulfill the criterion of privacy policy. It is a strict method but considering security concerns, it may suffice for the task. Lets take an example of runtime query enrichment.

Runtime Query Enrichment: The search of sensitive data can be constrained within predefined range of values and thereby it is restrained from direct querying by the end user. The user query is enriched automatically by the system before fetching the sensitive data within the predefined range of values. For example, Blood Pressure (BP) is a sensitive data. Assume that BP level from 90 to 120 is a range of a healthy person and 120 to 140 is a range of a non-healthy person. Therefore, the person falling in the range of BP level 120 to 140 may not like to disclose their BP levels to all other people and thereby, they would like to have a user access control on their BP level readings. In this case, the database column that holds the BP level readings is kept hidden from the users. And the user query is enriched at the runtime to invoke the data of the persons who fall in the range of BP level of 90 to 120. Consider the user query is "select * from person". Then this query is enriched through query modification as "select * from person where BPLevel >= 90 and BPLevel <=120".

6. **Audit Logging:** A log should be maintained that which person with what rights has accessed the system. Along with that, a log should include queries that a user executed on the system. This log combined with new queries can be used to evaluate queries in order to prevent privacy policy violations.

7. **Access Control:** Patients should have control over their own sensitive healthcare data, a technique has been proposed in [10]. Its should be decided by them whether they want to share the information with hospitals for research purposes or not.

8. **Access Privilege**: Hospitals should be having the privilege that once the data has been kept with hospital for any patient they can share among other hospitals for the research and development purposes.

4 Discussions

The literature shows that some solutions in India have mentioned to take security and user privacy into consideration [5, 6]. But, their proposals are not in compliance with the standards proposed by MHFW or any other international act or standardized policy set like HIPAA or HITECH Act. Thus, their is a need for imposing very stringent security policies and procedures. Security issues such as authentication, availability, confidentiality, integrity, access control, data ownership, data protection policies, user profiles and standard model need to be taken into consideration for EHRs. Techniques like *k-anonymity* [7] and *L-diversity* [8] should be used to make data more private and anonymous to disable the inferences from the databases. Incorporating security measures and privacy preserving techniques, organizations will benefit from increased user confidence, convenience, and speed of access to information.

5   Conclusions

We surveyed the problem of security and privacy for various EHRs already implemented and under development in India. In India, there is no specific Care Delivery Organization(CDO) [20] who monitors and follow a particular generic service delivery model. CDOs are legal entities whose primary mission is the delivery of healthcare related products and services. Our findings also implicate that most of the current systems lack for proper security and privacy measures being taken for the system and the user information. It outlines some of the security and privacy measures (figure 2), approved by the MHFW for EHRs solutions in India. But, still no solution has been developed on the basis of those guidelines. However, an insistence on demonstrated factors including implementation become part of future mandatory privacy and security concerns for EHRs systems.

It has been proposed what to look for in a secure EHRs system for complying with the security and privacy concerns at each layer (figure 3). A simulation for the implementation of the proposed model has been presented through various techniques like demographic and electronic heath records separation, query evaluation and modification, audit logging, access control and access privilege.

A very high level of security and privacy is required for the front-end user application and the back-end database. Thus, in future we aim to conduct actual implementation of the proposed model for an EHRs system based on semantic interoperability standard (such as openEHR or HL7).

References

1. State of the Union. Address of William J. Clinton USA (January 19, 1999)
2. ISO/TS 13606, http://www.iso.org/iso/home/store/catalogue_tc/catalogue_detail.htm?csnumber=50121
3. HIPAA Health Privacy Rule Act, http://www.hhs.gov/ocr/privacy/
4. Top 10 Data Security Breaches in 2012, http://www.healthcarefinancenews.com/news/top-10-data-security-breaches-in-2012
5. E-Sushrut, http://www.cdacnoida.in/healthcare.asp
6. DIGHT: Distributed Infrastructure for Global eHr Technology, http://dight.sics.se/?q=node/3
7. Sweeney, L.: k-Anonymity: A model for protecting privacy. International Journal on Uncertainty,Fuzziness and Knowledge Based Systems (2002)
8. Machanavajjhala, A., Gehrke, J., Kifer, D.: L-diversity: Privacy beyond k-anonymity. In: Proceedings of the 22nd International Conference on Data Engineering, Atlanta, GA, USA, April 3-8 (2006)

9. Addas, R., Zhang, N.: Support Access to Distributed EHR's with Three levels of Identity Privacy Preservation. In: Proceedings of Sixth International Conference on Availability, Relaibility and Security, Vienna, Austria, August 22-26 (2011)
10. Huda, M.N., Yamada, S., Sonehara, N.: Privacy-aware access to patient-controlled Personal Health Records in emergency situations. In: Proceedings of Third International Conference on Pervaisve Health, London, UK, April 1-3 (2009)
11. Donelan, K., Miralles, P.D.: supra note 17, at 66 (2006)
12. Law 41/2002 of November 14, basic regulator of the patient autonomy and rights and obligations of clinical information and documentation matters. BOE 274, sec. 1, pp. 40126-40132 (November 14, 2002)
13. Law 15/1999 of December 13, of the Protection of Personal Data BOE 298, sec. 1, pp. 43088-43099 (December 13, 1999)
14. Eichelberg, M., Aden, T., Riesmeier, J., Dogac, A., Laleci, G.: A survey and analysis of Electronic Healthcare Record standards. ACM Comput. Surv. 37(4), 277–315 (2005)
15. The openEHR Foundation, http://www.openehr.org
16. Wong, E., Stonebraker, M.: Access control in a relational data base management system by query modification. ACM SIGMOD (1975)
17. Biskup, J., Bonatti, P.A.: Controlled Query Evaluation for Known Policies by Combining Lying and Refusal. Annals of Mathematics and Artificial Intelligence 40(1-2), 37–62 (2004)
18. Health Information Systems Programmme, http://hispindia.org/
19. Electronic Health Record Standards For India, http://blog.digmed.in/2013/09/22/e-h-r-standards-for-india-goi-report/
20. Adams, J., Bakalar, R., Boroch, M., Knecht, K., Mounib, E.L., Stuart, N.: Healthcare 2015 and Care Delivery", IBM (white paper) (2013), http://www-03.ibm.com/industries/ca/en/healthcare/files/hc2015_full_report_ver2.pdf
21. Personal Health Records Need a Comprehensive and consistent Privacy and Security Framework, CTR. FOR DEMOCRACY AND TECHNOLOGY (June 9, 2009), http://www.cdt.org/policy/personal-health-records-need-comprehensive-and-consistent-privacy-and-security-framework
22. Tejero, A.: Advances and current state of the security and privacy in Electronic Health Records: Survey from a social prospective. Journal of Medical Systems 36, 3019–3027 (2012)
23. For the Record: Protecting Electronic Health Information, Committee on on Maintaining Privacy and Security in Health Care Applications of the National Information Infrastructures, National Research Council (1997)
24. Power, D., Slaymaker, M., Politou, E., Simpson, A.: Protecting sensitive patient data via query modification. In: SAC 2005. ACM (March 2005)
25. Carter, M.: Intergarted electronic health records and patients privacy: possible benefits and real dangers. Medical Journal of Australia 172, 28–30 (2000)

Minimizing Wind Resistance of Vehicles with a Parallel Genetic Algorithm

Shizuka Takako, Yayoi Takemura, and Lothar M. Schmitt*

The University of Aizu, Aizuwakamatsu, 965-8480, Japan
L@LMSchmitt.de

Abstract. We use a simple genetic algorithm with distributed fitness evaluation to optimize the simulated aerodynamic performance of idealized three-dimensional models of vehicles (car, truck) in a virtual wind-tunnel. The genetic algorithm manipulates part of the collection of numerical constants (genotype of candidate solutions) which describe the shape of said vehicle-models. For evaluation of fitness, the genotypes in a small population of 16–30 candidate solutions are translated into virtual vehicle-models (phenotypes) inside a mesh of specified density. The vehicle-models' wind-resistance which defines the inverse-proportional fitness of the associated genotypes is computed via the software package ELMER in parallel. After an optimization procedure over 100 generations when saturation of fitness improvement is seemingly achieved, we obtain (1) an optimized simple car-model which (up to practically necessary modifications) can be seen, in principle, on today's roads and (2), surprisingly, an optimized design for the driver's cabin of a truck and an air-shield on top of the latter which appears to be a *new* design. Both shapes found by the genetic algorithm described in this work significantly outperform elementary geometric shapes such as triangular or parabolic shapes in a simulation of aerodynamic resistance. The application of the findings in this work could yield to significant energy savings in the transport sector in the future.

Keywords: Aerodynamics, Computational Fluid Dynamics, Genetic Algorithm, Optimization, Parallel Processing, Road Vehicles.

Introduction

The goal of the work reported here which is based upon [26, 32, 35] is to have a simple genetic algorithm [9, 37] with parallel fitness evaluation find shapes of virtual, idealized three-dimensional models of vehicles (cars, trucks) with optimized aerodynamic performance using a low-budget[1], but otherwise state-of-the-art network of LINUX workstations. Thus, we also demonstrate that through use of low-cost hardware and open-source, free software tools, the optimization of the

---

* Corresponding author.

[1] The authors wish to thank *The University of Aizu* for providing internal, competitive funding for this project in the magnitude of USD 10^4 or JY 10^6.

A. Madaan, S. Kikuchi, and S. Bhalla (Eds.): DNIS 2014, LNCS 8381, pp. 214–231, 2014.
© Springer International Publishing Switzerland 2014

aerodynamic or hydrodynamic performance of products by computer-means is economically feasible in many situations where, *e.g.*, external parts for vehicles such as rear mirrors, or machinery which involves the transfer of gazes and fluids are produced.

Genetic algorithms were introduced in the 1960's in [3] (*cf.* the historical summary in [19]) and later in work described in [13]. By now, such algorithms are a well-established artificial intelligence method to solve optimization problems. Genetic algorithms have proven themselves to be among the most rubust & effective and, consequently, most often employed evolutionary methods for optimization purposes and have practical applications in many fields (see, *e.g.*, the various GECCO conference proceedings over many years). In order to apply such a method, a given problem instance is reformulated as an optimization task in such a way that

  o candidate solutions are understood as elements in a finite set C of creatures in a model "world", and
  o a so-called fitness function $f : C \to \mathbf{R}^+$ exists, which has to be maximized.

We note at this point that the fitness function f may even be population-dependent [25, sec. 2.6]. However, a population-dependent (modification of a given) fitness function needs to preserve a "natural" order in order to be used in an optimization procedure. Such a condition is clearly satisfied by, *e.g.*, rank or a linear rescaling depending upon the population (see sec. refsubsec:selection).

The genetic algorithm then evolves elements in C^s (usually $s \in 2\mathbf{N}$) called populations which, in most cases, is based upon the cyclic application of three major operations: mutation, crossover and fitness-value based, probabilistic selection. In most application settings, the number of elements in the finite set C is very large which prohibits a complete search of C. Genetic algorithms provide a probabilistic way to conduct a directed search in C in order to maximize f. These stochastic algorithms are particularly useful in applications for problem instances with irregular[2] shape of the given fitness function f.

In the present work, elements in $C \subset \mathbf{R}^\ell$, $\ell \in \mathbf{N}$, (the genotypes) describe the shape of a simple vehicle-model (car, truck) which is placed in a virtual wind tunnel. The number of ℓ-tuples of pseudo real numbers in C in our limited setting exceeds 10^{60}. The fitness function is essentially the reciprocal wind resistance force at $40\,\mathrm{km/h}$ of a moving virtual vehicle (the phenotype) but is rescaled for every population in order to strongly favor good candidate solutions. Consequently, the genetic algorithm used in this work pursues a perturbed variation of a hill-climbing approach via selection combined with local multi-directional search via mutation-crossover (see sections 1.3,1.4,1.7 for details).

In this work, we have used a so-called *simple* genetic algorithm [9, 37] where all principal operations stay constant over the course of the algorithm,

---

[2] By *irregular*, we mean here that the underlying fitness-function does not allow a rather obvious and computationally simple mathematical description. Thus, a genetic algorithm is often considered the method of choice, if for principal, timely and/or economic reasons, an efficient mathematical description of the underlying fitness function cannot be found for a problem instance in practice.

i.e., parameters such as the mutation rate stay constant over time. See [23–25] for a discussion of genetic algorithms where such parameters vary over time. The simple genetic algorithms in this work use a small population of 16–20 individuals with 10–15 real-valued genes, roulette-wheel fitness-proportional selection, single-cutpoint crossover with crossover rate 0.3 as well as multiple-spot mutation with rather large mutation rate 0.3 per creature and mutation rate 0.3 per gene for a mutated creature. The latter settings for mutation are in accordance with theoretical results in [25, Lem. 1.3.1, Prop. 2.6.1, Thm. 3.1.1] and [27, sec. 6] and experiments which are described in section 1.6.

In order to compute the fitness of candidate solutions and, thus, the wind resistance force of a moving vehicle or a stationary vehicle in a virtual wind tunnel, the Navier-Stokes equation [17, p. 45, eq. 15.6] needs to be solved for a region of space around the vehicle. The Navier-Stokes equation describes the motion of viscous fluid substances such as liquids and gases. In engineering practice, the Navier-Stokes equation is approximately solved by the finite difference method [46], the finite element method [20, 47], wavelet methods [2, 10, 14, 30] and others. In this work, the finite element method is implicitly used by employing the ELMER software package [40–42].

The finite element method is a numerical technique for finding approximate solutions of partial differential equations given as boundary value problem. Essentially (see [20, 47] for examples), the problem is approximately represented in a finite dimensional vector space and then transformed into a matrix equation which is solved by standard methods [42]. In our work, we have simply accepted and used the methods underlying the ELMER software package described in [4, 5, 42] as well as the methods described in [8] for meshing.

All in all, we obtain a significant improvement in aerodynamic performance, *i.e.*, 10%–20% reduction of wind-resistance force, for the optimized simple vehicle-models considered here compared to vehicle-models of elementary shape (*e.g.*, triangular roof). See the Conclusion for a detailed listing of results as well as a perspective of further research and testing planned for this on-going project.

In what follows, we shall present most of the methods used in this research by following the detailed description of the underlying genetic algorithm and its phases. Every major section of the description of the underlying genetic algorithm is usually devided into two parts, the description of the setup for the optimization of the car, and the description of the setup for the optimization of the truck.

1   Setup of the Genetic Algorithm Optimization

1.1   Genetic Algorithms and Optimization

As mentioned before, in this work, we have used a genetic algorithm with parallel fitness evaluation to optimize the shape of the idealized vehicles in regard to wind resistance. The wind resistance force on the vehicle model (a creature, or candidate solution in our genetic algorithm) which is placed in a virtual wind tunnel defines the reciprocal fitness function of a creature *cf.* sec. 1.7.

We note that our genetic algorithm uses a mixing phase consisting of mutation (see sec. 1.3), crossover (see sec. 1.4) and normalization (see sec. 1.5). The latter is usually not used in the setup of regular simple genetic algorithms, but is necessary here to forcefully keep the creatures or candidate solutions in a well-defined, limited domain.

1.2   Creatures and Populations

In our genetic algorithm's code, creatures are defined as lists of ℓ pseudo-real numbers describing the shape of the vehicle. The population is then a list of s creatures as described above.

A theoretical framework for using pseudo-real numbers to encode candidate solutions in a genetic algorithm can be found in [23, 25]. Using pseudo-real numbers instead of a binary encoding is quite common in genetic algorithm practice. See, *e.g.*, [12] for an approach using real-encoded genetic algorithms in a related research endeavor.

1.2.1   Creatures in the Simple Car Optimization

It is assumed that a vehicle has fixed length l=4.4m and width w=1.7m. The vehicle is seen from the side and divided into $n = \ell - 1$ segments of equal length. In our experiments, we set $n = 9$. A gene is a number in $[0.1, 1.5]$ and describes the height of one of the vehicle's segments in meters at length 0, $l/n, 2l/n, \cdots, l$. In any interval $[(\nu - 1)l/n, \nu l/n], 1 \leq \nu \leq n$, the height of the car is a linear function. The only requirement for the car's shape is that at least one of the ℓ genes equals 1.5m. The population size is set to $s = 20 > \ell$ in accordance with [23], Thm. 7.2.3, Ex. 7.3]. Figures 1, 2, 3 and 4 show four possible creatures in our setting.

1.2.2   Creatures in the Truck (Driver's Roof) Optimization

Creatures (genotypes) are used to define the shape (phenotype) of the roof on top of the driver's cabin. The roof is symmetric in width-direction with respect to the center line of the truck. The roof is seen from the top and divided into $n_w - 1 = 4$ intervals along width direction and $n_l - 1 = 4$ intervals along length direction. This yields a $n_w \cdot n_l$ height values matrix defining the shape of the roof. A creature is a list of $\ell = \lceil n_w/2 \rceil \cdot n_l$ height values describing the entries in "left half" of the matrix. The "right half" of the latter matrix is obtained by symmetry. For example, the genotype, *i.e.*, 5 × 3 matrix

$$((.6, .7, 1.), (.5, .6, .7), (.3, .4, .7), (.15, .25, .55), (.1, .2, .5))$$

corresponds to the phenotype shown in Fig. 6. The population size is set to $s = 30 > \ell$.

1.3   Mutation

Mutation implements random change in the genetic information of candidate solutions. Mutation is the canonical noise in a genetic algorithm which assures (weak) ergodicity of the underlying inhomogenous or homogeneous Markov

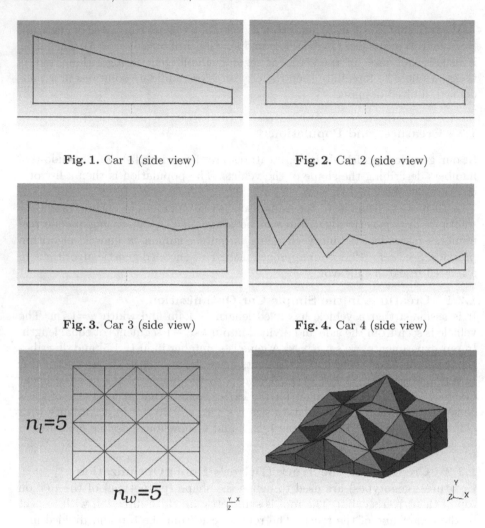

Fig. 1. Car 1 (side view)   **Fig. 2.** Car 2 (side view)

Fig. 3. Car 3 (side view)   **Fig. 4.** Car 4 (side view)

Fig. 5. Layout of the Truck Roof, top view   **Fig. 6.** Truck Roof, frontal view

chain. This, in turn, assures a unique limit-probability distribution over the set of all possible states of the genetic algorithm which desirably has the most weight near and in populations containing globally optimal creatures. Proofs in regard to the latter statements can be found in [23, 25]. In our setting, mutation is executed as follows: (1) In a loop over the population, every creature in the population is mutated with probability μ_c. (2) If mutation is applied to a creature, every gene in the creature is mutated with probability μ_g. The procedure to mutate a single gene g_o into a new gene g_n is given by $g_n = r \cdot g_o$ where $r \in [0.9, 1.1]$ is a uniformly distributed random number. Thus, in our setting, we rather pursue a hill-climbing approach with the mutation operator, *i.e.*, a creature is only subject to a small change/deformation under mutation.

1.4   Crossover

Crossover implements the recombination of genetic information of candidate solutions. Crossover is understood as being the operation of a genetic algorithm which has the potential to combine good partial solutions (here shapes) to an overall good global solution of the problem instance. In our setting, crossover should ideally combine a good front of one vehicle with a good rear of another vehicle.

Crossover is in our setting applied in the following way to a population $(c_1, \ldots, c_s)$ of $s \in 2\mathbf{N}$ creatures: first, the crossover procedure pairs creatures in the population sequentially as (c_1, c_2), $(c_3, c_4), \ldots, (c_{s-1}, c_s)$. Then, crossover is applied to the individual pairs with probability χ, the crossover rate. The latter crossover operation on pairs is tailored to the specific geometries of the problem instance (car or truck), and is described in detail in sections 1.4.1 and 1.4.2 below.

Note that our crossover operators yield offspring identical to the parents for an identical pair of parents. This is a theoretical requirement in the analysis of genetic drift in [27, sec. 6], [23, sec. 6.5].

1.4.1   Crossover in the Simple Car Optimization

If crossover is applied to a pair of cars O_1 and O_2 (See figures 7,8), a cut-point at one of the segment boundaries is decided at random. First, a new car N_1 is made with the part of car O_1 left of the cut-point and the part of car O_2 right of the cut-point. A second new car N_2 is made similarly combining the remaining parts.

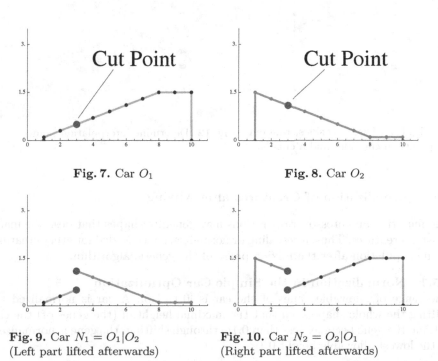

Fig. 7. Car O_1 **Fig. 8.** Car O_2

Fig. 9. Car $N_1 = O_1 | O_2$ **Fig. 10.** Car $N_2 = O_2 | O_1$
(Left part lifted afterwards)   (Right part lifted afterwards)

When parts of the cars $O_{1,2}$ are combined, there may be a gap (discontinuity) at the cut-point (see figures 9,10). We repair the gap by lifting the partial shape with lower value at the cut point by the length of the gap (see sec. 1.5).

1.4.2   Crossover in the Truck (Driver's Roof) Optimization

If crossover is applied to a pair of creatures, it is first randomly decided either to use the *"front and back"* or the *"left and right"* sides (edges) of the rectangle limiting the halved creature. Recall that creatures are representing the left half of the roof and the right half is determined by symmetry.

Then a random cut-line deviding the rectangle into two halfs is determined as follows: two random points on each of the selected edges (lines) are chosen and the cut-line that connects the two points is defined (see figure 11). The interpolation points of each column (or row) that are the closest points to the cut-line are calculated (see figure 12).

Crossover in the truck optimization exchanges all matrix values of the parents below (or left of) the cut-line. At the interpolation point, the gap is repaired by taking the average of the heights of the parents.

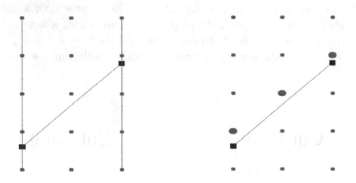

Fig. 11. Chose two sides, random points on the sides and a cut-line   **Fig. 12.** Determine interpolation points

1.5   Normalization of Creatures after Mixing

The mutation and crossover operations may generate shapes that describe inadmissible creatures. Thus, a rescaling or normalization is needed for every creature in the population after the mixing phase of the genetic algorithm.

1.5.1   Normalization in the Simple Car Optimization

The range of allowable genes of the car is [0.1, 1.5]. A car is normalized by shifting the whole shape such that the maximal height of (the genes of) the car is 1.5m. If a gene becomes less than 0.1m through shifting, the gene is normalized to the lowest allowable height 0.1m.

1.5.2 Normalization in the Truck (Driver's Roof) Optimization

Let h_c be the height of the trucks cabin. Let h_b be the height of the cargo section (*e.g.*, a loaded container) of the truck. If there are some genes out of allowable range $[0, \ 2(h_b - h_c)]$, the roof is normalized. The roof is normalized by multiplying all genes by a constant such that the maximal gene length is $2(h_b - h_c)$.

1.6 Preparatory Experiment on Mutation and Crossover Rate

In this project, we experimented with several combinations of mutation rate μ_c and crossover rate χ for $\mu_g = 0.3$. From the results of one particular set of trials (Fig. 13–16), we found that higher rates of crossover and mutation yielded a faster increase in top-fitness per generation. Accordingly, we used $\mu_c = \chi = 0.3$. According to [27, sec. 6], a very low mutation rate μ_c yields a genetic algorithm that is more likely to implement genetic drift (with rare larger jumps in the maximal fitness in the population). Having a larger mixing rate accelerates the convergence of the algorithm in accordance with [25, Lemma 1.3.1] in a probabilistic sense.

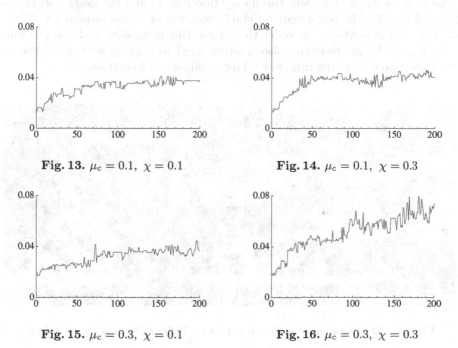

Fig. 13. $\mu_c = 0.1, \ \chi = 0.1$

Fig. 14. $\mu_c = 0.1, \ \chi = 0.3$

Fig. 15. $\mu_c = 0.3, \ \chi = 0.1$

Fig. 16. $\mu_c = 0.3, \ \chi = 0.3$

1.7 Fitness Selection

Fitness selection models reproductive success of adapted organisms in their environment. The genetic algorithm used in this work employs roulette-wheel selection as described in [9, p. 11]. In general, the raw fitness value $f_r(c)$ of a creature c is defined as follows:

$$f_r(c) = \frac{1}{|\text{non-zero air resistance force in z-direction}|}$$

for computation of the air resistance force. The z-direction is frontal vertical direction in our experiments while y denotes height and x denotes width. The computation of the air resistance force uses the software packages GMESH (see sec. 2.1) and ELMER (see sec. 2.2).

Following [9, p. 77], the fitness is linearly rescaled such that the ratio between the best and worst creature in a population equals 100:1. This is done to accelerate convergence in our hill-climbing approach. Thus, we have a population-dependent fitness function as considered in [25, sec. 2.6].

1.8   Fitness Evaluation and Eliminating the Effect of Tourbulence

In simulating the Navier-Stokes equation, the simulation software (ELMER) obtains flow solutions that show turbulence (see Fig. 17 and the lower part of [43, p. 28, Fig. 5.2]). In particular, no stable solution of a flow similar to [43], p. 23, Fig. 4.2] is achieved. In order to counter this instability and the resulting variation of the air resistance force, we decided to take an average of the air resistance force in z-direction over a larger number of iterations.

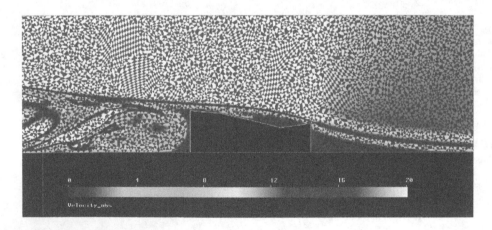

Fig. 17. 2D simulation of airflow behind Car 3 (figure 3) showing turbulence

1.8.1   Turbulence Effects for the Simple Car

After 11 iterations ELMER's simulation shows only instability in a turbulent region behind the car and there is no significant change in the average resistance force value after taking into account iterations 11–110 (see Tbl. 1).

Table 1. Average air resistance force [N] on 4 cars (see Fig. 1–4) over generation $11-\nu$, $\nu \in \{110, 130\}$

generations	car1	car2	car3	car4
11–130	−63.4	−65.6	−124.9	−92.2
11–110	−63.1	−65.5	−124.9	−92.3

1.8.2 Turbulence Effects for the Truck

After 11 iterations ELMER's simulation shows only instability in a turbulent region behind the truck and there is no significant change in the average resistance force value after taking into account iterations 11–60 (see Tbl. 2).

Table 2. Average air resistance force [N] on 4 random trucks over generation $11-\nu$, $\nu \in \{60, 130\}$

generations	Truck 1	Truck 2	Truck 3	Truck 4
11–130	−181.0	−224.4	−243.3	−256.6
11–60	−181.3	−224.2	−243.0	−255.0

1.8.3 Distance between the Car and the Ground

In our series of preparatory experiments, we found that if there is no gap between the car and the ground, the air in the front of the car becomes an air cushion and virtually changes the shape of the car and the result significantly. Table 3 shows that with no gaps, Car 3 becomes a relatively good creature because of the air cushion in front of it (see Fig. 17). Therefore, we keep 0.3m between the simple car and the ground and keep 0.5m between the truck and the ground.

Table 3. Resistance force [N] depending upon the distance d[m] between car and ground (see Fig. 1–4)

d	Car 1	Car 2	Car 3	Car 4
0.0	−179.9	−125.5	−163.6	−275.7
0.3	−217.8	−201.4	−260.8	−291.7

1.8.4 Ommission of Wheels in the Simulation

For reason of efficiency and feasibility, we simplified the setting by not including wheels in our models. First, observe that the wheels are relatively far away from structures of interest to us (top/roof-shape). Secondly, observe that this omission applies uniformly to all creatures and thus, should not change our results significantly.

1.9 Executing the Genetic Algorithm

The genetic algorithm employed in this work essentially runs on a single computer, henceforth, called the *main computer*. However in our work, computation of the fitness function is very time consuming (90 mins. for one fitness value). Therefore, the calculation of the wind resistance force which represents reciprocal fitness is executed in parallel on a family of *branch computers* that are connected with the main computer via Internet and were mounted on the main computer using Network File System [51]. Every branch computer shares a specifically named directory with the main computer. In the shared directories, files can be written, edited and removed by both the main computer and the respective branch computer. In this way, the subtasks of fitness evaluation were communicated from the main computer to the branch computers, and the fitness values of creatures in the population retrieved in reverse. This work was carried out on an inhomogeneous network of UNIX work stations (PCs under Linux Mandrake and Apple Mac under Mac OS X).

2   Solving the Navier-Stokes Equation with GMESH and ELMER

In order to compute the wind resistance force, GMESH is used to generate a mesh around the vehicle model filling the virtual wind tunnel. The latter mesh is then used as input to ELMER in order to solve the Navier-Stokes equation for the particular geometry described by the mesh. This computation yields a pressure field [40, sec. 2] around the vehicle model in the virtual wind tunnel. The wind resistance force on the vehicle is then computed from the latter pressure field [40, sec. 33].

2.1   Making the Mesh Pertaining to a Specific Creature

We use GMESH to create mesh files which ELMER can process. To start the process, we define an outer box (see sec. 2.2.2). These are the dimensions of the encompassing virtual wind tunnel. The dimensions of the virtual wind tunnel are chosen large compared to the dimensions of the vehicle to avoid interference of the wall with the vehicle. Thus, our simulation should approximately describe a vehicle driving in the open. The car is represented as a carved-out shape in the mesh.

GMESH is an automatic three-dimensional finite element mesh generator with built-in pre-processing and post-processing facilities. This software package is free and open source. For a detailed overview on GMESH, see [8]. See, in particular, [8, p. 3], for details how the mesh is generated in a bottom-up fashion. See [6, 7, 29, 38] for a description of the underlying meshing algorithms. Figure 18 shows the result of an application of GMESH.

2.1.1   Characteristic Length λ
For our simulations, the sizes of underlying matrices depend upon the number of points in the mesh. This number depends upon the so-called characteristic length λ. Following [8, p. 54] every mesh generating step is constrained by the characteristic length field. The size of mesh elements is computed by linearly

Fig. 18. Part of a mesh around a simple car

interpolating the characteristic lengths on the initial mesh. In our application, this is specified by characteristic lengths associated with points in the geometry [8, sec. 4.2.1]. If λ is chosen too short, computational time and computer memory requirements increase. On the other hand, if λ is too large, the simulation will ignore fine details of the shape of the vehicle and the air flow turbulence (see Fig. 17 and Tbl. 4). With the coarse mesh (λ=0.8), Car 4 is the best creature while Car 2 is the best creature with the fine mesh (λ=0.1). Based upon these preparatory experiments, we decided to set λ=0.1m close to the car since, in this case 4λ is the width of the variable car section, λ=0.2m on the ground for reason of continuity, and λ=0.5m on the ceiling of the tunnel on the simple car simulation. On the truck simulation, we set λ=0.05m close to the roof of the truck, λ=0.2m close to the trucks cabin, the cargo section and on the ground and λ=0.5m on the ceiling of the tunnel.

Table 4. Air resistance force [N] on 4 cars (see Fig. 1–4) depending upon λ[m]

λ	Car 1	Car 2	Car 3	Car 4
0.1	−218.1	−215.9	−256.9	−286.4
0.8	−55.6	−162.2	−65.1	−37.9

2.1.2   Preparatory Experiment on the Random Factor of Meshing
Since GMESH uses random numbers to generate a mesh, the meshes pertaining to the same creature but generated at distinct times usually differ. Therefore, we confirmed that simulations with ELMER for different meshes for one particular creature yield similar values for the wind resistance force. Tbl. 5 shows that this is true in general as Car1–3 show very similar results and Car 4 turns out to be rather bad in each case.

Table 5. ELMER's simulation based upon different meshes with similar characteristics (see Fig. 1–4)

pc	Car1	Car2	Car3	Car4
linux (1)	−179.9	−163.6	−125.5	−275.7
linux (2)	−180.4	−169.0	−142.1	−304.7
mac (1)	−185.5	−167.1	−126.0	−232.7
mac (2)	−203.3	−167.1	−123.3	−249.0

2.2 Solving the Navier-Stokes Equation with ELMER

The wind resistance force on the vehicle is calculated using ELMER. It is a finite element software package for the solution of partial differential equations *cf.* [41, p. 1]. This software package is free and open source. ELMER is divided into several separate executables [41, p. 4] that can be used independently. The executables relevant for our project are `ElmerGrid` and `ElmerSolver`. `ElmerGrid` serves as a filter which transforms the mesh file obtained from GMESH into an ELMER-internal format. `ElmerSolver` has a solution model for the Navier-Stokes equation among many physical models. We use ELMER's material constants for air at room temperature as follows:

density $= 1.205[\text{g/m}^3]$,   heat capacity $= 1.05 \cdot 10^3 [\text{J/(kg·m)}]$,

heat conductivity $= 0.0257[\text{W/(K·m)}]$,   reference pressure $= 1.0 \cdot 10^5 [\text{Pa}]$,

sound speed $= 343.0[\text{m/s}]$,   specific heat ratio $= 1.4$,   viscosity $= 16.7 \cdot 10^{-6} [\text{Ns/m}^2]$.

We assume that the vehicle moves in z-direction. The air enters the tunnel uniformly with 12.0m/s= 43.2km/h in (−z)-direction. At the surface of the ground, the air can't move up and down in y-direction. At the surface of the vehicle, the air can't move [43, Tutorial 4]. Everywhere else, the air can move freely. The Navier-Stokes equation is solved in the above setting. Then, the wind resistance force on the vehicle surfaces is calculated based on the flow solution and the pressure field.

2.2.1 Preparatory Experiment on Non-linear Iteration Methods

In the air flow simulation, we initially experimented with the use of both the generally slower Picard method [48] and the Newton method [52] implemented in ELMER. However, we found that use of the Newton method yielded unstable/undefined solutions to the Navier-Stokes equation. As a measure to assure a reliable feedback from the branch computers to the main computer, we decided to use the Picard iteration method exclusively.

2.2.2 Preparatory Experiment on Box Size

We experimented with the dimensions of the box surrounding the car model since a smaller box means smaller matrices as input for ELMER and shorter simulation time. However, we found by visual inspection of ELMER's results that the surrounding box should extend about two car lengths behind the car. It seems that this space is needed for the simulation to "remember" (keep trace of) turbulence behind the car. After two car lengths the turbulent air flow behind the car seems to calm down. Any shorter box yielded numerical instabilities in our simulation, *i.e.*, flow calculation of ELMER diverged. For this reason, we decided to extend the simulation box for two cars length behind the car. For similar reasons, the box extends one car length in front of the car. A similar setting is used for the simulations related to the truck model.

3   Experimental Results for the Car Optimization

The genetic algorithm was run with 10 branch computers, $\ell = 10$ and $s = 20$ (*cf.* sec. 1.2.1), $\mu_c = \mu_g = \chi = 0.3$ (*cf.* sec. 1.6), distance $0.3m$ between the car and the ground (*cf.* sec. 1.8.3). It took about 20 days to process 100 generations. The best creature of the 100th generation is shown in Fig. 19. Recall, that the only requirement for the car was its overall height 1.5m at some point on the roof.

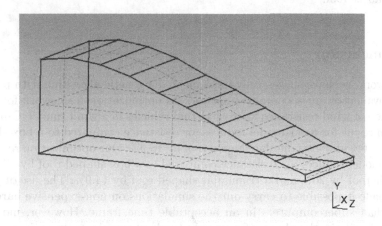

Fig. 19. Best car shape in a simulation

4   Experimental Results for the Truck Optimization

The genetic algorithm was run with 16 branch computers, $\ell = 15$ and $s = 16$ (*cf.* sec. 1.2.2), $\mu_c = \mu_g = \chi = 0.3$ (*cf.* sec. 1.6), distance $0.5m$ between the car and the ground (*cf.* sec. 1.8.3). The genetic algorithm was started with a population having 50% triangular and 50% rectangular roofs. It took about 20 days to process 70 generations. Then, another genetic algorithm was started with the best creature of the above genetic algorithm. The genetic algorithm was run with 17 faster branch computers, $\ell = 15$ and $s = 30$ (*cf.* sec. 1.2.2), $\mu_c = \mu_g = \chi = 0.3$ (*cf.* sec. 1.6), distance $0.5m$ between the car and the ground (*cf.* sec. 1.8.3). It took about 10 days to process 100 generations. The best creature of the 100th generation is shown in Fig. 20. It beats the shown parabolic roof in regard to wind resistance by reducing the air resistance by about 20%.

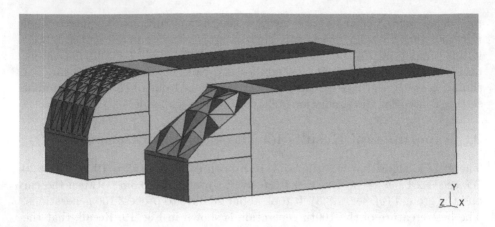

Fig. 20. Foreground: Best truck shape in a simulation. Background: Reference shape with parabolic roof.

5   Conclusion

In this work, we have successfully implemented a genetic algorithm with parallel fitness evaluation on a computer network as optimization framework for a 3D simulation of wind resistance force of virtual vehicles in a wind tunnel. Our optimization result for a car model reduces air resistance compared to a box-shaped roof by 74.0% and a triangular shaped roof by 18.9%. Our optimization result for a truck model reduces air resistance compared to a box-shaped roof by 47.8%, a parabolic roof by 20% and a triangular shaped roof by 14.9%. The use of parallelism made it possible to carry out the simulations on non-expensive hardware rather than supercomputers in an acceptable time frame. However, modeling realistic details in the shape of a vehicle (such as wheels while optimizing the roof) seems currently out of reach for our low-end computer network. In our experimental setting, the matrices arising in the linearized problem description obtain via finite element method from the Navier-Stokes equation seem to grow too large to be handled effectively. Further research in regard to the latter and several other aspects (optimization of choice of parameters and generation of meshes) of this work will have to be carried out in the future.

References

1. Batchelor, G.K.: An introduction to fluid dynamics. Cambridge University Press (1967)
2. Bratteli, O., Jørgensen, P.: Wavelets Through a Looking Glass: The World of the Spectrum. Birkhauser (2002)
3. Bremermann, H.J., Rogson, J., Salaff, S.: Global Properties of Evolution Processes. In: Pattee, H.H., et al. (eds.) Natural Automata and Useful Simulations, pp. 3–42. Spartan Books, Washington, DC (1966)

4. Franca, L.P., Frey, S.L., Hughes, T.J.R.: Stabilized finite element methods: I. Application to the advective-diffusive model. Computer Methods in Applied Mechanics and Engineering 95, 253–276 (1992)
5. Franca, L.P., Frey, S.L., Hughes, T.J.R.: Stabilized finite element methods: II. The incompressible Navier-Stokes equations. Computer Methods in Applied Mechanics and Engineering 99, 209–233 (1992)
6. Geuzaine, C., Remacle, J.: Gmsh: a three-dimensional finite element mesh generator with built-in pre- and post-processing facilities. Int. J. Numer. Meth. Eng., 1–24 (2009)
7. George, P.-L., Frey, P.: Mesh generation, Hermes (2000)
8. Geuzaine, C., Remacle, J.: Gmsh Reference Manual. University of Liége (2009), http://www.geuz.org/gmsh/doc/texinfo/gmsh.html1
9. Goldberg, D.E.: Genetic Algorithms in search, Optimization & GoldbergGAbook-Machine Learning. Addison-Wesley (1989)
10. Greibel, M., Koster, F.: Adaptive Wavelet Solvers for the Unsteady Incompressible Navier-Stokes Equations. Advanced Mathematical Theories in Fluid Mechanics, 67–118 (2000)
11. Hilbert, R., Baron, G.J., Thevenin, D.: Multi-objective shape optimization of a heat exchanger using parallel genetic algorithms. Int. J. of Heat and Mass Transfer 49, 2567–2577 (2006)
12. Huang, L., LeBeau, R.P., Hauser, T.: Application of genetic algorithm to two -jet control system on NACA 0012 air foil Comp. In: Fluid Dynamics 2004, Part VI, pp. 349–354. Springer (2004)
13. Holland, J.H.: Adaptation in Natural and Artificial Systems. University of Michigan Press (1975); Extended new ed. MIT Press (1992)
14. Jørgensen, P., Treadway, B.: Analysis and probability: wavelets, signals, fractals Springer (2006)
15. Kernighan, B.W., Ritchie, D.: The C Programming Language, 2nd edn. Prentice Hall (1988)
16. Kimura, H.: Mathematica, Shuwa System, Tokyo (2001)
17. Landau, L.D., Lifshitz, E.M.: Fluid Mechanics, 2nd edn. Butterworth Heinemann (2007)
18. Lang, S.: Analysis I. Addison-Wesley (1969)
19. Mühlenbein, H.: In: E. Aarts, L. Lenstra (eds.) Local Search in Combinatorial Optimization, 47–54. Wiley (1997)
20. Nikishkov, G.P.: Intro. to the FEM (2007), http://web-ext.u-aizu.ac.jp/~niki/feminstr/introfem/introfem.html (August 28, 2008)
21. Rebay, S.: Efficient unstructured mesh generation by means of Delaunay triangulation and Bowyer-Watson algorithm. J. Comput. Phys. 106 (1993)
22. Rudin, W.: Real and Complex Analysis. McGraw-Hill (1974)
23. Schmitt, L.M.: Theory of Genetic Algorithms. Theor. Comp. Sci. 259, 1–61 (2001)
24. Schmitt, L.M.: Asymptotic Convergence of Scaled Genetic Algorithms to Global Optima —A gentle introduction to the theory. In: Menon, A. (ed.) Frontiers of Evolutionary Computation. Kluwer Series in Evolutionary Computation, pp. 178–221 (2003)
25. Schmitt, L.M.: Theory of Genetic Algorithms II: models for genetic operators over the string-tensor representation of populations and convergence to global optima for arbitrary fitness function under scaling. Theor. Comp. Sci. 310, 181–231 (2004)
26. Schmitt, L.M.: Notes on Computational Fluid Dynamics and Genetic Algorithm Optimization (2008, 2009)

27. Schmitt, L.M., Nehaniv, C.L.: The Linear Geometry of Genetic Operators with Applications to the Analysis of Genetic Drift and Genetic Algorithms using Tournament Selection. In: Nehaniv, C.L. (ed.) Mathematical and Computational Biology: Computational Morphogenesis, Hierarchical Complexity and Digital Evolution. Lectures on Math. in the Life Sci., vol. 26, pp. 147–166. AMS (1999)
28. Schmitt, L.M., Nehaniv, C.L., Fujii, R.H.: Linear Analysis of Genetic Algorithms. Theor. Comp. Sci. 200, 101–134 (1998)
29. Schoeberl, J.: Netgen, an advancing front 2d/3d-mesh generator based on abstract rules. Visual. Sci., 41–52 (1997)
30. Schneider, K., Kevlahan, N.K.-R., Farge, M.: Comparison of an Adaptive Wavelet Method and Nonlinearly Filtered Pseudospectral Methods for Two-Dimensional Turbulence Theor. Comp. Fluid Dynamics 9, 191–206 (1997)
31. Takako, S.: Installing Gmsh, Notes and Manual, The University of Aizu, 1–14 (2008)
32. Takako, S.: Minimizing Wind Resistance of Vehicles with a Parallel Genetic Algorithm. Part I: Theory. B.S.-Thesis, The University of Aizu (2010)
33. Takako, S.: Finite Element Method (Poster), The University of Aizu (2008)
34. Takemura, Y.: Elmer Software Package, Notes and Manual, The University of Aizu, 1–42 (2008)
35. Takemura, Y.: Minimizing Wind Resistance of Vehicles witha Parallel Genetic Algorithm. Part II: Experiments. B.S.-Thesis, The University of Aizu (2010)
36. Thomas, J.W.: Numerical Partial Differential Equations. Springer (1995)
37. Vose, M.D.: The simple Genetic Algorithm: Foundations and Theory. MIT Press (1999)
38. Weatherill, N.P.: The integrity of geometrical boundaries in the two-dimensional Delaunay triangulation. Commun. Appl. Numer. Methods 6(2), 101–119 (1990)
39. Clay Mathematics Institute: Navier-Stokes Equations, http://www.claymath.org/millennium/Navier-Stokes_Equations/ (November 17, 2009)
40. CSC–IT Center for Science, Elmer Models Manual (September 8, 2009), http://www.csc.fi/english/pages/elmer/documentation
41. CSC–IT Center for Science, Overview of Elmer (September 8, 2009), http://www.csc.fi/english/pages/elmer/documentation
42. CSC–IT Center for Science. ElmerSolver Manual, http://www.csc.fi/english/pages/elmer/documentation (September 8, 2009)
43. CSC–IT Center for Science, Elmer Tutorials (September 22, 2009), http://www.csc.fi/english/pages/elmer/documentation
44. Wikipedia: Derivation of the Navier-Stokes Equations (September 16, 2009), http://en.wikipedia.org/wiki/Derivation_of_the_Navier-Stokes_equations
45. Wikipedia: Divergence theorem (November 25, 2009), http://en.wikipedia.org/wiki/Divergence_theorem
46. Wikipedia: Finite difference method (September 18, 2009), http://en.wikipedia.org/wiki/Finite_difference_method
47. Wikipedia: Finite Element Method (September 16, 2009), http://en.wikipedia.org/wiki/Finite_element_method
48. Wikipedia: Fixed Point iteration, http://en.wikipedia.org/wiki/Picard_iteration (September 16, 2009)
49. Wikipedia: Material derivative (November 23, 2009), http://en.wikipedia.org/wiki/Convective_derivative
50. Wikipedia: Navier-Stokes Equations (September 16, 2009), http://en.wikipedia.org/wiki/Navier-Stokes_equations

51. Wikipedia: Network file system (November 26, 2009),
    http://en.wikipedia.org/wiki/Network_file_system
52. Wikipedia: Newton's method (September 16, 2009),
    http://en.wikipedia.org/wiki/Newton's_method
53. Wikipedia: PicardLindelöf theorem (September 8, 2009),
    http://en.wikipedia.org/wiki/Picard%E2%80%93Lindel%C3%B6f_theorem
54. Wikipedia: Reynolds transport theorem (November 23, 2009),
    http://en.wikipedia.org/wiki/Reynolds_transport_theorem

Kaguya Moon Mission Data Repository: New Query Language Interface for Locating GIS Objects

Hiroki Nakamura, Subhash Bhalla, Junya Terazono, and Wanming Chu

Graduate Department of Computer and Information Systems,
University of Aizu, Fukushima, Japan
{m5161134,bhalla,terazono,w-chu}@u-aizu.ac.jp

Abstract. The observation data gathered by a luanr probe "Kaguya" [14], is now being officially published by Japan Aerospace Exploration Agency (JAXA) through the web site. However, this web site does not have the lunar location and name based search function, thus we can not retrieve the data by location name or by feature type. Therefore, we developed the lunar feature/name based Kaguya data search system. In addition, this system can be used by the simple keyword input, making use of Kaguya data. However, at present, only a geometric image product "TCOrtho_MAP" [15] is available among the 88 Kaguya products [11]. We plan to store, adopt all remaining products in a similar way, but the importing operation is not completely automated yet. Also it is hard to say that this system is highly optimized for dealing with Kaguya data. Because the base of this system is another support system called the Moon Seeker, it is for a lunar feature searching. Since finish of the Kaguya mission, some processed data, such as 3D map [4], are still generated by Kaguya's observation data. Hence we hope to continue the enrichment of this system for further promotion of these demands for Kaguya data.

Keywords: Kaguya mission, Moon Seeker, Kaguya Data Archive, GIS, Lunar features, Seasar2, OpenLayers.

1 Introduction

1.1 Kaguya Mission Outline

In the middle of 2007, the Japan Aerospace Exploration Agency (JAXA) launched a lunar orbiter which was named "Kaguya" (English name: SELENE) [14]. This orbiter was injected to the lunar orbit on October 2007, and had observed many kind of moon data for about one year and a half until JAXA maneuvered the orbiter to impact it onto the moon surface in June 2009. The main scientific objectives of the Kaguya mission were to study the origins of the moon and its geological evolution, obtain information about the lunar surface environment for future lunar utilization [13]. In addition, Kaguya orbiter also performed a near moon space study of radio science, plasma, electromagnetic field and high-energy particles by

A. Madaan, S. Kikuchi, and S. Bhalla (Eds.): DNIS 2014, LNCS 8381, pp. 232–255, 2014.

its instruments. The former lunar orbiters missions did not obtain such kind of scientific data, thus Kaguya's data is being analyzed by lunar scientists around the world even today.

1.2 Kaguya Data Archive

Observational data obtained by Kaguya is processed to reduce noise, reformatting location data and arrange to proper data format. These processed data is called as "Product", are produced from one instrument by different product level or instrument function. These Kaguya mission products are now stored and managed by the website named "Kaguya Data Archive" [10], offered by JAXA. In the Kaguya Data Archive, we can search and download a specific data file by the specified search condition such as product ID, lat/long, time range and area selection. These search conditions are given from file's metadata(e.g., observation date, instrument name, vesion number, etc.). The metadata is associated with individual single observed file, and this metadata is called "Catalog Information" in this archive. Therefore, the detail observation file search with Catalog Information is available in this archive. In contrast, the Kaguya Data Archive does not have a search function that uses lunar features or location/object name.

1.3 Other Lunar Data Archives

As similar data archives by other lunar orbiters, NASA provides the "Lunar Mapping and Modeling Project (LMMP)" [16] and the "Lunar Orbital Data Explorer" [30]. Both websites have the lunar features(e.g., crater, mountain, and sea) and keyword search function as shown in Fig. 1. Not only the Catalog Information search, but these website can find the particular observation data by using both methods, the search by using the instruments approach and the geography approach. For example, by typing the keyword into a input form, the LMMP returns the result that are matched from both location name and Catalog Information. The data generated by Kaguya mission is not supported by a suitable search system.

1.4 Moon Seeker

Some lunar location based search systems which do not contain the observation data searching have been developed. For example, University of Aizu, developed the lunar location search site "Moon Seeker" [21]. This website can be used for searching by Query-By-Example method using lunar feature, lon/lat, diameter, ethnicity, and others. Fig. 2 is a search form page of the Moon Seeker. Also, by using previous search result as additional search element, the Moon Seeker enable high-level compound retrieval like "Find location which is within 100km from previous search result's location" or "Find location which is overlapped with previous search result's location". In addition, we can visually check the

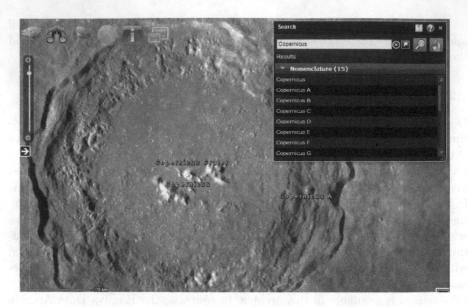

Fig. 1. Screenshot of LMMP with keyword search

Moon Seeker

Help

Find objects their conditions are,

Feature Type	Crater ⏷ ?		
Name	tycho		
Direction	— ⏷		
Latitude	Min:	Degrees, Max:	Degrees
Longitude	Min:	Degrees, Max:	Degrees
Diameter	Min: 50	km, Max:	km
Ethnicity	— ⏷		
Continent			

Fig. 2. Search condition page of the Moon Seeker

retrieved location by use of Google Earth API [5]. This function displays the search result as a marker onto the lunar Geographic Information System (GIS). For this lunar feature search system, it enables us to do more searching, compared to the "location find function" which comes with above instance of NASA's data archive(LMMP and LODE).

1.5 Modifing the Moon Seeker for Kaguya Mission

Common specification are used between data archives of the orbiter and the lunar location search systems. Both systems deal with latitude/longitude for handling athe observation data and lunar location. Furthermore, the Moon Seeker has the circumradius of each lunar location, and Kaguya's Catalog Information has the observational range of all four corner's lat/long value. This means not only the center point of location, but also those can mark the area of target. Therefore, by extracting the Catalog Information from each Kaguya's observation data and storing that into the same database with the Moon Seeker, it permits displaying the Kaguya's observed data which is overlapped with retrieved locations from the Moon Seeker. By combining the Moon Seeker and the Catalog Information of Kaguya Product, we can find the specific Kaguya's observation data based on lunar features. In this study, we aimed to combine the Moon Seeker and the Catalog Information of Kaguya's data to make the new Kaguya archive as above. This is for easing strain on the searching from huge amount of Product files, to support the researcher who study the Kaguya data, and improve the utilization of Kaguya data.

2 Method

To develop the new mashup service with "Kaguya Data Archive" with "Moon Seeker", we need to combine both systems.

2.1 About Kaguya Data

Kaguya Data Structure. The lunar explorer "Kaguya" was equiped with 15 instruments. Some (refined data constellation from raw observation) data, it is called "Product", is produced by every single instrument. The number of Products which were created by Kaguya mission amounts to 88 types [11]. In these each Product, it has a number of archived files, those have a different time range or observation area. Basically each archived file consist of image or numerical primary data with metadate(named "Catalog Information") and thumbnail image. As seen above, the structure of Kaguya data is composed of an instruments, it has some Products, and a Product has many archived files which is lumped primary data and ancillary data together. This structure is known as Planetary Data System (PDS) [17]. PDS was invented by NASA, and this is very popular and widely used format in any planetary science projects. Kaguya's data is also created according to this PDS format (Fig. 3).

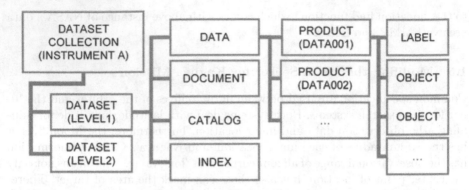

Fig. 3. PDS file structure format http://www.isas.ac.jp/docs/PLAINnews/188_contents/188_1.html

Specification of Kaguya Data Archive. As described in section 1, Kaguya Data Archive is the official website to manage the Product that is based on observation data by lunar explorer "Kaguya". This website can narrow the search by instruments type, product type, observation time range, observation area with longitude/latitude or version number. Additionally, this web site has the lunar surface GIS for selecting the specific area on moon by drag and drop. Each retrieved result (specific file) by above search function have a web page, it is called "Catalog detail page" (Fig. 4). The Catalog detail page enable us to check the metadata of retrieved file without downloading the actual data. Also the thumbnail image is embedded in this Catalog detail page. Each element of this metadata in the Catalog detail page is commonly used in all other products' file and include the following information.

- File name
- Product name, Product ID
- Data size
- Thumbnail image and URL
- Longitude/Latitude of four corners of the observation area
- Observation time
- Comment
- etc

2.2 The Moon Seeker

Present Specification. Moon Seeker, as exlpained in section 1, is the lunar feature/nomenclature search site. This web site is implemented by Seasar2(Java Framework) [29], PostgreSQL [27] and Apache Tomcat [22], and the lunar surface data of Google Earth API is being used for visualization of search result. Also, the stored lunar feature data in database is acquired from the web site "Gazetteer of Planetary Nomenclature" [7], this is managed by International

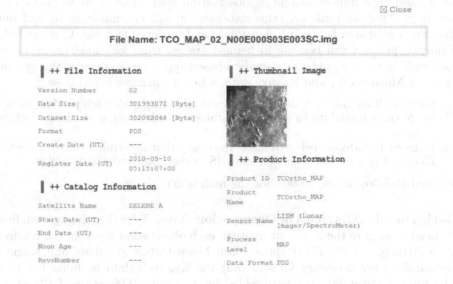

Fig. 4. Example of the Catalog detail page

Astronomical Union (IAU), U.S. Geological Survey (USGS) and NASA. By inputting the feature's type, name, longitude/latitude, diameter, etc to this form on screen, we can find the satisfied features. It is also important function to remember that this system can add the search result features for new search condition as additional objects. It is possible to drill down search by means of these operators "overlap", "disjoin", "near", "contain" and previous search result. By using this function, for example, such kind of advanced queries are realized that "Find all features within 100km from Mountain A" or "Find all craters which is contained in a crater with more than 500km radius crater". This "searching with previous search result" function is not implemented to any other lunar feature search site, such as Lunar Mapping and Modeling Project or Lunar Orbital Data Explorer.

2.3 Outline of the Method

For searching the Kaguya's observation data based on the lunar feature, We need two database tables. One is the data about lunar features. This every single lunar feature(e.g., crater, mountain) has some parameter such as feature type, longitude/latitude, area, ethnicity, etc. These feature data amounts more than 9,000 [8] but already downloaded from IAU web site. Also this data is already stored in our laboratory server's database and used by the Moon Seeker for searching. Another one is the database table of each Kaguya's observation data. Each generated datasets by Kaguya's instruments is given the metadata.

In the metadata, it contains a mission name, instrument name, observation time range, and most important one is "observation area". Therefore, to create and put both database table on same database, it will be enable us to find out the crossovered area between Kaguya data and retrieved features. Consequently Kaguya's product will become to being retrieved from location/feature based approach, instead of products' metadata based approach. To develop this system by using Moon Seeker and Kaguya data, a broad workflow is as follow.

1. Gather all metadata which are attached with Kaguya's each product data
2. Store these metadata to the same database which also has the lunar feature data
3. Improve the Moon Seeker to make the query that determine the overlapping
4. Display the matched result on the GIS in the improved Moon Seeker.

we need to follow above 4 steps for the realization.

Gathering the Metadata of Observation Data. To realize this system, first we need to acquire the X,Y coordinate of each observation file from the Kaguya Data Archive. As described above, both Moon feature database and Kaguya file database are necessary for searching the Kaguya's data by lunar features. The lunar feature data list is provided by a website "Gazetteer of Planetary Nomenclature" which is described in section 2.2. We already downloaded this data as csv file, put into the database and this database is used by the Moon Seeker. Therefore, this time we only gather the Kaguya's observation data. The information of each obseravtion data is also provided as a html file in the archive for checking without data download. By downloading these html files one-by-one with web crawler, prepare the metadata which has X,Y coordinate as same amount as the number of files.

Storing the Metadata to the Database. After downloading each html file, we need to analyze the HTML code structure for extracting actual the metadata. However, the total file number is over thousands, thus it is impossible to extract one-by-one manually. Therefore, we develop a Java code that loads each html file, then read that code line-by-line and return only the matched data as a SQL query. This program will load all html files based on a prepared file path list, then extract the metadata, handle a missing value, and finally output a "INSERT INTO" query as a text file. At this time, this Java program also add the new rectangle value column for representing the observation area. Because the metadata only has the X,Y coordinate of four corner, and it does not have the geometric rectangle data. By executing this program, we get all metadata with X,Y coordinate as SQL query in a specific product. Finally, execute this SQL query script on the PostgreSQL database which is used by the Moon Seeker, complete the process of metadata extracting.

Implement the Overlap Decision. Next, consider about how to implement the overlap judgement with the existing lunar feature's geometric data and prepared metadata. PostgreSQL offers the geometric data type, thus we use this to express the area of lunar features or observed data. This geometric data has some operators such as "overlap", "contain", "distance", "congruent" [9]. In this case, we check "Is the retrieved lunar feature pictured by a specific observation area or not", thus use the operator "&&" for the overlap operation. The return value of this overlap operator is boolean type, and will return "true" if certain feature A and B are overlapped, or "false" if not. Hence, it is possible to find the specific Kaguya data which is accorded with the retrieved lunar features by execute the SQL query kind a "The lunar feature area && The observation area of specific product WHERE lunar feature.ID = retrieved lunar feature.ID". This overlap judgement behavior is expressed as Fig. 5.

Fig. 5. Overlap judgement between a crater and observation images

Matching the Kaguya Data and the Moon Seeker. Finally, to execute previous SQL query via actual web site, we need to modify the code of the Moon Seeker. The current Moon Seeker is for finding the lunar features from a keyword or X,Y coordinate range, and showing this result on the GIS. Now we modify this GIS page to add a pull-down menu of Kaguya Product list and a submit button of Product search. Then create a new search result page for display appropriate Kaguya data by using the submitted product code from previous button and hidden SQL executing page. Thereby the screen transition "feature search → search result → show up on GIS → back to search" has 2 more steps "product selection → show up Kaguya data on GIS" at the bottom. In this way, based on the Moon Seeker developed by Mr.Tsunokake, we implemented the system which enable us to find the Kaguya data from the new aspect of "lunar features" by mixing 2 databases.

3   Environment and Implementation

3.1   Environment

The Moon Seeker, as explained at section 1 and 2, becomes the basement of this Kaguya data search system. Thus, because we modify the Moon Seeker to add the Kaguya data search function, definitely the environment of this system is also based on the environment of the Moon Seeker. To develop this system, the developing environment is integrated by Intergrated Development Environment, Eclipse [23]. It arises from the Moon Seeker uses JavaServer Pages (JSP) [20] for dynamic page generating, "Seasar2" framework(described below) uses Java, and this Seasar2 plug-in is optimized for Eclipse. The Moon Seeker was constructed by the Java framework "Seasar2", and it is now running on the Apache Tomcat under Linux server. This Seasar2 framework, especially for using the O/R Mapper called "S2JDBC" [28], the smooth connection between this website and PostgreSQL is realized. Specifically, there is a mechanism to do automatic variable generation for search condition form from the database, and keep it after page transition. Also the Seasar2 has prepared search functions such as "full-list acquisition", "condition entry acquisition", "number of items acquisition" for easy query making when the system executes a SQL query based on the conditions of search form entry by users. In addition, to the extent of improving the Moon Seeker's environment, we changed the GIS from Google Earth API to OpenLayers [25] [2] for search result visualization. The main factor of changing is that the map display of Google Earth API supports only the Windows and Mac OS X. This means there is no way to check the search result by Ubuntu, this is one of the Linux distribution, which is used as the develop environment of this system. Not only the Ubuntu, but also Linux is used by Andriod. This OS is now widely used for smartphones and tablets, and increasing an increase of Android users is expected. Therefore we adopted the OpenLayers for adapting the accesses from these mobile devices. Furthermore, the OpenLayers can handle the wide variety of map layer other than Google Earth. By using OpenLayers, we can use a lot of map layers deveoped and provided by Web Map Service (WMS) [18], ArcGIS [1], MapServer [24]. Therefore the system easily adapts any map layers based on the lunar missions other than the Kaguya mission. Fig. 6 is the overall view of this system.

3.2   Implementation Details

In this paper, use the product which is called "TCOrtho_MAP" [15] to explain a series of steps for importing the Product to the Moon Seeker. The Product "TCOrtho_Map" is the image data which was taken by the instrument "Terrain Camera (TC)" [12] and then mosaic processed data. An Image data is

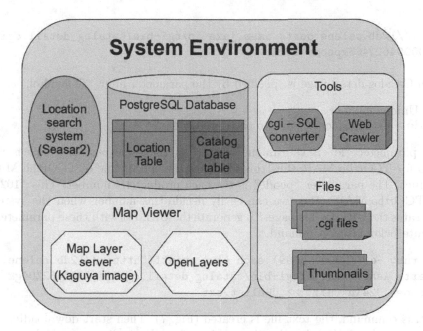

Fig. 6. Overall system environment of this system

the popular data in the Kaguya product, and also the total number of this data is 7,200. That is not too many compare with other image product, thus TCOrtho_MAP is the suitable example for the explanation. Therefore this paper uses the TCOrtho_MAP as an example, however the step is totally same as following step to import all other products.

Creating a Crawler. To develop this system, firstly we need to do is gathering the metadata which are attached with Kaguya's each product data. Fundamentally the matadata is called as "Catalog Information File", and it is stored as .cfg file(text data) with the main data and the thumbnail image in each compressed dataset. However, the total amount of all data provided by the Kaguya Data Archive is more than 22TB, and the number of all data is over 1.8 million [11]. Hence download all file for getting each metadata is impossible by temporally and capacitatively reason. Therefore we try to extract the same metadata from Catalog detail pages in Kaguya Data Archive(explained in section 1.2 and 2.1) instead of downloading each Catalog Information File with the heavy main data. The size of every Catalog detail page is less than 10KB, it is good for downloading. For that, we need to create the web page crawler to download every Catalog detail page in Kaguya Data Archive automatically. As explained in section 2.1, Kaguya Data Archive has the Catalog detail page and the thumbnail image. Thereby, firstly we create the each wep page and thumbnail's URL list by using Ruby [33]. The Catalog detail page's URL structure is as follow.

http://l2db.selene.darts.isas.jaxa.jp/cgi-bin/catalog_detail.cgi?
id=00004622468&pcode=1024

Each Catalog detail page is specified by the parameter after "?" symbol.

id : Unique number
pcode : Fixed value assigned to each products

The parameter "id" is the unique number in the same product. This is the
URL for TCOrtho_MAP, thus this id means the specific file of TCOrtho_MAP
product. The parameter "pcode" is the each product ID number. This "1024"
is "TCOrtho_MAP", thus we can easily decide this number when the system
generates the URL to this page. To generate the URL list with these parameters,
execute below Ruby command.

```
% ruby -e '(2368..9999).each {|n| printf("https://l2db.selene.
darts.isas.jaxa.jp/cgi-bin/catalog_detail.cgi?id=0000462%04g&
pcode=1024\n",n)}' >> URLlist.txt
```

By this command, the text file is created (Fig. 7). Then start downloading all
specific product's Catalog detail page by using Wget [3] command based on
generated URL list. Also we have to set the cookie information to the HTTP
header, by "–header" option of Wget. Because the Kaguya Data Archive uses
cookie for timeout handling or Japanese-English page switching. Fig. 8 is the
result after executing Wget command with Catalog detail page URL list.

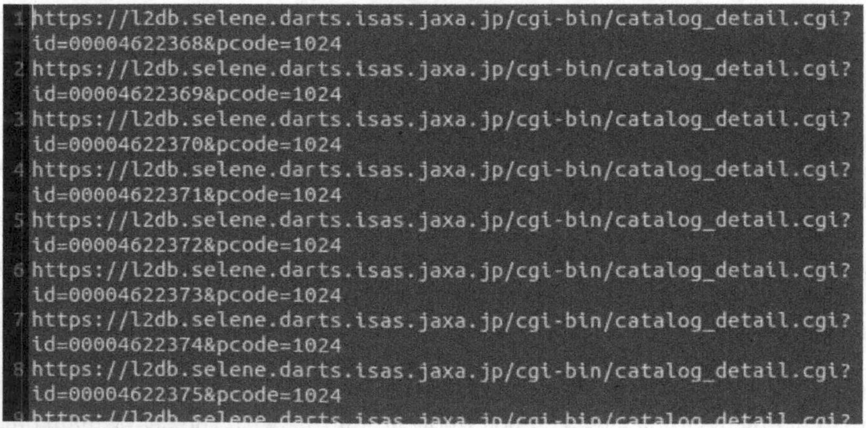

Fig. 7. Generated URL list by Ruby command

Fig. 8. Downloaded all catalog detail pages

Creating a Database Table. Before extract the metadata from downloaded Catalog detail pages, we need to create the database table. Because without the database structure, we can not extract and create the SQL query(INSERT INTO command) from Catalog detail pages. To store the products' data generated by Kaguya, prepare two tables. One is product list, another is all catalog information(metadata). Below figure is the all database table which will be used by this system, include the Moon Seeker. In this time, we newly created the "product_code" table and the "catalog_detail" table, put on the right side of Fig. 9. For creating the table, firstly we need to execute the "CREATE TABLE" command in the PostgreSQL. The defined data type in each table become as Table 1 and Table 2. The data type which is used the catalog_detail table is based on the Product Format Description book(P.117-126) [15], offered by JAXA. The format of each data type is described in this book, then the corresponding PostgreSQL data type was selected and adopted for new database table. Also there is "pcode" integer type column in the catalog_detail table. This catalog_detail table's pcode value is corresponded to the pcode column which is stored in the product_code table, and connected as the foreign key. After creating both table by CREATE TABLE command, then run INSERT INTO query to create tuples of metadata. First, it is about the product list table. Each Product have the specific product name and the product code. To acquire the list of all product name and product code, we use the source code(HTML code) of Kaguya Data Archive. When we use the product search in the Kaguya Data Archive, there is a product selection function by pull-down menu. Therefore, open the Kaguya Data Archive's source code via web browser, and copy the content of <option> tag where is nested

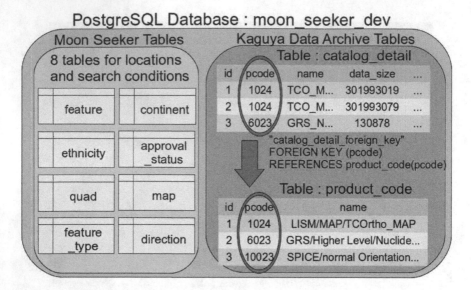

Fig. 9. Whole database structure of this system

Table 1. Structure of the product_code table

Column	Type	Modifiers
id	integer	not null default nextval('product_code_id_seq'::regclass)
pcode	integer	not null
name	text	not null

Indexes:
    "product_code_pkey" PRIMARY KEY, btree (id)
    "product_code_pcode_key" UNIQUE CONSTRAINT, btree (pcode)
Referenced by:
    TABLE "catalog_detail" CONSTRAINT "catalog_detail_foreign_key"
    FOREIGN KEY (pcode) REFERENCES product_code(pcode)

in the <select> tag for pull-down menu, we get the correspondence table of the product name and the product number (Fig. 10). After that, use the replace function of the text editor for clearing the unnecessary HTML tag and make a comma-sepalated paird text file of product code and name. Additionally, we prepaFred a Java program which load this text file, add the sequential number for the primary key and output as the SQL query. By this program, finally we get the SQL query list as Fig. 11. Next, explain about the catalog_detail table. To create the catalog_detail table's tuple as INSERT INTO queries of SQL, we needs to analyse the HTML source structure of catalog detail pages and extract only the required data. For that, we have to read every single catalog detail page what are downloaded by the previous section's crawler. Therefore, we developed

Table 2. Structure of the catalog_detail table

Column	Type	Modifiers
id	integer	not null default nextval('catalog_detail_id_seq'::regclass)
pcode	integer	not null
name	text	not null
version_number	integer	
data_size	integer	
⋮		
upper_left_longitude	real	not null
upper_left_latitude	real	not null
upper_right_longitude	real	not null
upper_right_latitude	real	not null
lower_left_longitude	real	not null
lower_left_latitude	real	not null
lower_right_longitude	real	not null
lower_right_latitude	real	not null
data_area	box	not null
⋮		

Indexes:
    "catalog_detail_pkey" PRIMARY KEY, btree (id)
Foreign-key constraints:
    "catalog_detail_foreign_key"
    FOREIGN KEY (pcode) REFERENCES product_code(pcode)

another Java code for doing these steps automatically. The catalog_detail table's tuple making is further illustrated in next section with the parser for HTML source analysis.

3.3   Creating the Parser

After downloading the catalog detail pages from the Kaguya Data Archive, we need to do the extracting of the metadata from each file. The extension of this catalog detail page is ".cgi", it represents and stands for "Common Gateway Interface" [31], and virtually this is html file. The total number of downloaded catalog detail pages by the crawler is 7,200 files in this "TCOrtho_MAP" case. However this will definitely depend on Product type. To extract the required metadata by reading the html code line-by-line for each of the 7,200 html files, the HTML parser was developed by use of Java. This parser was named "Cgi-ToSql.java", and firstly it loads the prepared cgi file list of catalog detail pages by using readFileList() function. Then this program reads the file name in order from top by BufferedReader() function, and loads all files one-by-one by read-FileList() function again. After load a catalog detail page, the parser discerns the type of HTML tags by using "HTMLEditorKit" [19] library for Java, and

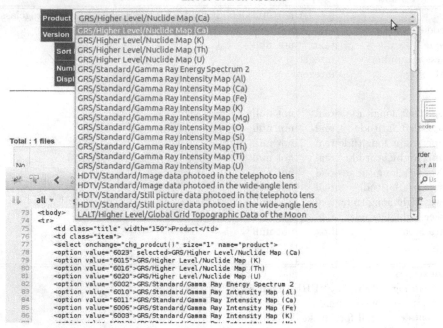

Fig. 10. Product list in the HTML source code of Kaguya Data Archive

```
31 INSERT INTO product_code VALUES(31,1020,'LISM/L2B/SP_Level2B1');
32 INSERT INTO product_code VALUES(32,1021,'LISM/L2B/SP_Level2B2');
33 INSERT INTO product_code VALUES(33,1017,'LISM/L2C/MI-NIR_Level2C2');
34 INSERT INTO product_code VALUES(34,1014,'LISM/L2C/MI-VIS_Level2C2');
35 INSERT INTO product_code VALUES(35,1011,'LISM/L3D/DTM_TCOrtho');
36 INSERT INTO product_code VALUES(36,1025,'LISM/MAP/DTM_MAP');
37 INSERT INTO product_code VALUES(37,1026,'LISM/MAP/MI_MAP');
38 INSERT INTO product_code VALUES(38,1024,'LISM/MAP/TCOrtho_MAP');
39 INSERT INTO product_code VALUES(39,1010,'LISM/MAP/TC_Evening_MAP');
40 INSERT INTO product_code VALUES(40,1009,'LISM/MAP/TC_Morning_MAP');
41 INSERT INTO product_code VALUES(41,9005,'LMAG/Higher Level/1D electrical
   structure');
42 INSERT INTO product_code VALUES(42,9001,'LMAG/Higher Level/Magnetic anom
43 INSERT INTO product_code VALUES(43,9002,'LMAG/Higher Level/Magnetic anom
44 INSERT INTO product_code VALUES(44,9003,'LMAG/Standard/Magnetic field t
45 INSERT INTO product_code VALUES(45,9008,'LMAG/Standard/Magnetic field t
   (Option)');
46 INSERT INTO product_code VALUES(46,8004,'LRS/Higher Level/Subsurface ge
```

Fig. 11. A part of INSERT INTO queries for product list

processes depending on that. This is the actual Java code steps to invoke the parser for HTML tags in the CgiToSql.java.

```
ParserDelegator parser = new ParserDelegator();
HTMLEditorKit.ParserCallback mycallback = new MyCallBack();
parser.parse(new FileReader(file), mycallback, true);
```

For each loaded cgi file, this Java program reads the HTML code line-by-line, and checks the used HTML tag to change the response. Then the CgiToSql.java extracts only the metadata from cgi files and finally converts the metadata to the SQL's INSERT INTO query and output as text. Specifically the all metadata is nested by html's <td> tag, thus the program outputs only the line when loads <td> tag (Fig. 12). However, the line nested by <td> tag is not only the metadata, but also there is the label information. Because of that, the program is improved to load only the value of data for the boolean value flag. The loaded metadata's value will be processed such as remove unnecessary characters, replace the "NULL" if it is empty, and get dumped as a part of SQL query. Also when the program gets four corners X,Y coordinate of the observation area, the program adds new PostgreSQL's "box" type value to express the rectangle. Because the observation area is expressed as this box type, the system can make the overlap judgement by using the lunar surface features in other table. Then finally one INSERT INTO query is generated from one catalog detail pages by above steps. In this case "TCOrtho_MAP", as already described in above, there is 7,200 catalog page files. Thus also 7,200 INSERT INTO queries are output by this CgiToSql.java. The html template which is used by all catalog detail page are same regardless of the instrument of Product. It means, the structure of all

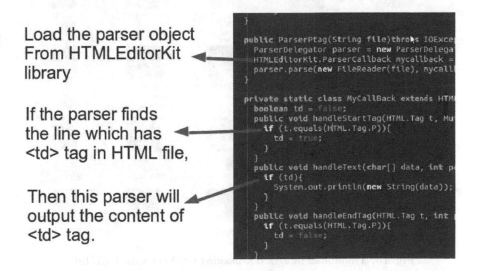

Fig. 12. Explanation about the code of CgiToSql.java

catalog detail page's html code is same, and we can extract the metadata for all Products with only this CgiToSql.java program. In addition, by using this parser to get the file name from the catalog detail page, also download the thumbnails which is associated with each pages. The URL format of thumbnail image is as below.

```
https://l2db...load_datafile.cgi?t=t&f=TCO_MAP_02_N00E000S03E00
3SC&p=1024&s=UklHbTZJdHAgaXx2fjNAPmRKTz1K
```

- parameter **f** means product file name
- parameter **p** means product code(1024=TCOrtho_MAP)
- parameter **s** means session ID(cookie)

Of these URL's trailing parameters, the values of "f" and "p" vary with each file. Thus try to make the URL list of thumbnail URL dynamically by extracting the file name and product code with above parser. Then download all thumbnail images by using Wget command same as when download all catalog detail pages. Finally we get the whole thumbnail images in our local storage (Fig. 13). This images will be used when the system display the result of specific Kaguya data onto GIS.

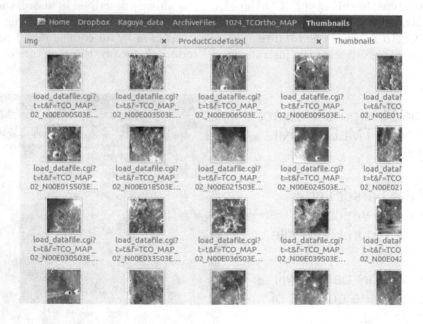

Fig. 13. Thumbnail images downloaded by Wget with URL list

3.4   The Overlap Judgement

To judge the overlap between lunar features and observation area from the database, use the geometric data operator what is described at section 2. In this case, to check the areas are overlapped or not, connect both geometic variables by "&&" operator and use this return value. Before creating the query via web site, first we test the overlap judgement by executing the SQL query directly under PostgreSQL terminal. Access to the Moon Seeker's database and then execute below SQL query.

```
SELECT catalog_detail.name, catalog_detail.data_area
FROM feature, catalog_detail WHERE feature.id = 8748
AND catalog_detail.pcode = 1024 AND feature.circle_area &&
circle(catalog_detail.data_area) = true
ORDER BY feature.name ASC;
```

This is the query to get the column and output the file name and this area which return "true" in regard to the condition "lunar feature's ID is 8748" and "product code is 1024". Here each feature.id = 8748 and catalog_detail.pcode = 1024 means the "Tycho" crater and the "TCOrtho_MAP". Tycho is a characteristic and well-known crater among the lunar features because it has over 1500km ray in the full moon. Reduced to its simplest terms, this query searchs the pictures which only contain the Tycho crater, from all 7,200 TCOrtho_MAP files and output in ascending order of file name. As a result, 44 TCOrtho_MAP files are hit and output as Fig. 14. In the actual service, it is not always true that the retrieved feature become only one item, thus all features which apply to conditions will be output as a list. Hence, this system put the ID numbers of feature group to an array, then execute above SQL query with shifting feature.id by for loop. Finally, by a combination of all obtained catalog_name.id(or name), the system can do the product search for all features which was retrieved by user's search condition.

3.5   Matching the Code

To execute this SQL query from actual website, we need to prepare some new dynamic web pages. In current version, the website shows the results of lunar feature searching on Google Earth lunar map, then returns to the search page again. In this time, there is a need to create new product search menu bar and submit button into above GIS page. To select and submit the specific product by this pull-down menu, the hidden judgement page will be called for checking the overlap between retrieved features and selected product. Then the SQL query will be executed, and overlapped file ID will be returned. After that, this hidden SQL page passes these values of overlapped file ID (to the new GIS page). A new GIS page will receive that ID value, and render the polygons [26] [2] of each observation area based on that ID. This means the new GIS webpage will create a vector layer which has the geometric polygon objects to express each observation area on OpenLayers.

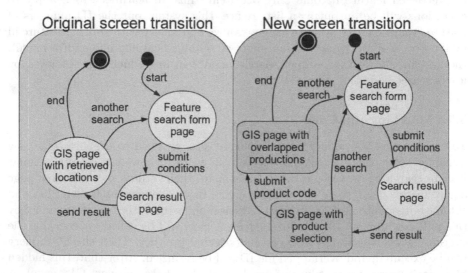

```
⊗ ⊖ ⊙    m5161134@X60: ~
         name                    |              data_area
---------------------------------+----------------------------------
TCO_MAP_02_N18E000N15E003SC.img  | (2.99976,18),(0,15.00024)
TCO_MAP_02_N15E000N12E003SC.img  | (2.99976,15),(0,12.00024)
TCO_MAP_02_N12E000N09E003SC.img  | (2.99976,12),(0,9.00024)
TCO_MAP_02_N12E003N09E006SC.img  | (5.99976,12),(3,9.00024)
TCO_MAP_02_N09E000N06E003SC.img  | (2.99976,9),(0,6.00024)
TCO_MAP_02_N09E003N06E006SC.img  | (5.99976,9),(3,6.00024)
TCO_MAP_02_N06E000N03E003SC.img  | (2.99976,6),(0,3.00024)
TCO_MAP_02_N06E003N03E006SC.img  | (5.99976,6),(3,3.00024)
TCO_MAP_02_N03E000N00E003SC.img  | (2.99976,3),(0,0.00024)
TCO_MAP_02_N03E003N00E006SC.img  | (5.99976,3),(3,0.00024)
TCO_MAP_02_N03E006N00E009SC.img  | (8.99976,3),(6,0.00024)
TCO_MAP_02_N00E000S03E003SC.img  | (2.99976,0),(0,-2.99976)
TCO_MAP_02_N00E003S03E006SC.img  | (5.99976,0),(3,-2.99976)
TCO_MAP_02_N00E006S03E009SC.img  | (8.99976,0),(6,-2.99976)
TCO_MAP_02_S03E000S06E003SC.img  | (2.99976,-3),(0,-5.99976)
TCO_MAP_02_S03E003S06E006SC.img  | (5.99976,-3),(3,-5.99976)
TCO_MAP_02_S03E006S06E009SC.img  | (8.99976,-3),(6,-5.99976)
TCO_MAP_02_S06E000S09E003SC.img  | (2.99976,-6),(0,-8.99976)
TCO_MAP_02_S06E003S09E006SC.img  | (5.99976,-6),(3,-8.99976)
TCO_MAP_02_S06E006S09E009SC.img  | (8.99976,-6),(6,-8.99976)
TCO_MAP_02_S09E000S12E003SC.img  | (2.99976,-9),(0,-11.99976)
```

Fig. 14. The result of overlap judgement between Tycho crater and TCOrtho_MAP

Fig. 15. Comparison of Original Moon Seeker and Product search version

By adding the new screen transition from previous system, the system is adapted to the Kaguya data. The actual screen transition figure as Fig. 15. Also, to add the downloaded thumbnail images for each OpenLayers objects, and display that by clicking, the user can check how the actual observation data is. Furthermore, also those objects have the URL to access to the real observation data stored in the Kaguya Data Archive. Each URL for real data stored in the Kaguya Data Archive is defined by the file name, thus we can generate such downloading URL by file name list stored in our local database. As mentioned above, we add the new single GIS page into the Moon Seeker, and put 3 below elements into the that GIS for adapting the Kaguya data.

1. Generate the rectangle objects to display the observation area of overlapped Kaguya data
2. Display the thumbnail image
3. Generate the URL Link to the real observation file

4   Comparison of Results

The system has been implemented for the Kaguya product search with lunar features. In this section, explain the result of this developed system with an example. First, find the location with condition "feature type = crater" and "name = Copernicus" by using Moon Seeker's feature search funtion. Then select the Copernicus crater which is appeared in the result list. After that, we can find the metadata of this Copernicus crater such as X,Y coordinate, diameter, ethnicity, etc. Also in this page, there is a link to the GIS page for checking the retrieved feature in a visual way. The GIS displays the location of Copernicus crater, and this page also has the pull-down meun bar for Kaguya Products. Select the one of the Kaguya Product, click the submit button, and finally we access to the new GIS page. This page shows the specific Kaguya data includes the Copernicus, as Fig. 16. Fig. 16 is the example of Kaguya file searching. As seen in Fig. 16, the retrieved Kaguya data are expressed as the yellow box on the GIS. That is how to find the specified Kaguya data from user selected product.

5   Discussion

The main problem is that, the PostgreSQL's geometric data operation supports only the same data type. The lunar feature mainly consist of (say) craters, thus it uses the "circle" data type. On the other hand, the observation data uses the "box" data type, the data for rectangles. Therefore, when the system judges the overlap between circle data type and box data type, the box type data is casted to the circle type. Therefore, it has the possibility to return the wrong judgement in such case (Fig. 17). Hence, we need to consider some new method to accurately judge the overlap between circle type and box type, not to be forced to typecast to the circle type. Also, we should consider the holding method of search result. At the moment, to replicate the specific search result(e.g., previous result or

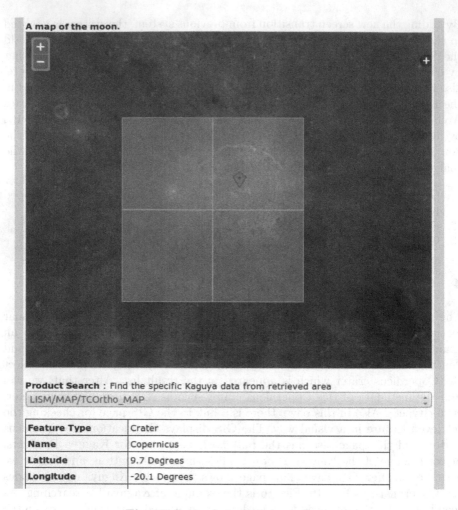

Fig. 16. Screenshot of overlapped area

other users result), we need to search again with same condition. For every search result, we should prepare the parmanent link by using the random string. Using to this parmanent link, we can share the search result more effectively.

One more problem, also we need to consider about continuous observation data updating. For the lunar feature, this is the natural objects and we do not need to update frequency. Howeve about the Products, it have the potential to add or update another product by lunar scientists. The response to this expected products in future, a total automation of all catalog detail page and thumbnails downloading, SQL generating, executing the SQL can be positive. If this automation program is realized, all we need is just run this program, thus anyone can update the metadata up-to-date. Hence, this automatic maintenance program is necessary for continuous operation.

Expected Result   Actual Result

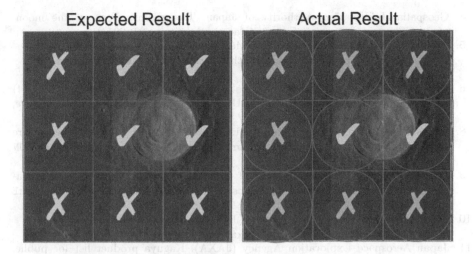

Fig. 17. Miss judgement by circle casted observation areas

6   Summary and Conclusions

The lunar feature based Kaguya data searching has become possible. Previously we could do the Kaguya data search with only time range, coordinate, or version number. Thus if the user remembers the specific lunar surface location as the place-name instead of X,Y coordinate, it is hard task to find that location. Thus we can find the lunar location or data in the familiar way using earth GIS(e.g., Google Maps [6], Yahoo! Maps [32]) operation. However, it is still the case that the system can search only a portion of Products, it is not adapted for all products. The X,Y coordinate searching which uses the lunar features as a condition, can be adapted to any other Kaguya mission data not only the map data. Therefore, we aim to upgrade this system to integrate lunar search system through connecting it to the other lunar mission's data.

With the advancement of technology, today we can explore the universe with more high sensitive sensors. On top of everything, on the bounty of the generalization of Web, now we can freely touch the space probe's observation data even individually. This exponentially-increased exploration data by high-performance instruments, is currently not being managed and stored effectively. This study aims to improve the user interface using free software tools.

References

1. Environmental Systems Research Institute, Inc. (ESRI). ArcGIS,
   http://www.arcgis.com/ (accessed January 27, 2014)
2. Hazzard, E.: OpenLayers 2.10 Beginner's Guide. Packt Publishing (March 2011)
3. Free Software Foundation, Inc. GNU Wget, http://www.gnu.org/software/wget/
   (accessed January 27, 2014)

4. Geospatial Information Authority of Japan. Topographic maps of the moon, http://gisstar.gsi.go.jp/selene/ (accessed January 27, 2014)

5. Google Inc. Google Earth API - Sky, Mars, and Moon, https://developers.google.com/earth/documentation/sky_mars_moon (accessed January 27, 2014)

6. Google Inc. Google Maps, https://maps.google.com/ (accessed January 27, 2014)

7. International Astronomical Union (IAU). Gazetteer of Planetary Nomenclature, http://planetarynames.wr.usgs.gov/ (accessed January 27, 2014)

8. International Astronomical Union (IAU). Gazetteer of Planetary Nomenclature - The Moon, http://planetarynames.wr.usgs.gov/SearchResults?target=MOON (accessed January 27, 2014)

9. IThe PostgreSQL Global Development Group. Geometric Functions and Operators. http://www.postgresql.org/docs/9.3/static/functions-geometry.html (accessed January 27, 2014)

10. Japan Aerospace Exploration Agency (JAXA). Kaguya Data Archive, https://l2db.selene.darts.isas.jaxa.jp/ (accessed January 27, 2014)

11. Japan Aerospace Exploration Agency (JAXA). Kaguya product list for public, http://l2db.selene.darts.isas.jaxa.jp/help/en/KAGUYA_product_list_public_en.pdf

12. Japan Aerospace Exploration Agency (JAXA). KAGUYA (SELENE) - Mission Instruments - LISM [TC, MI, SP], http://www.kaguya.jaxa.jp/en/equipment/tc_e.htm (accessed January 27, 2014)

13. Japan Aerospace Exploration Agency (JAXA). KAGUYA (SELENE) - Mission Profile, http://www.kaguya.jaxa.jp/en/profile/index.htm (accessed January 27, 2014)

14. Japan Aerospace Exploration Agency (JAXA). Kaguya (SELENE, Selenological and Engineering Explorer), http://www.kaguya.jaxa.jp/index_e.htm (accessed January 27, 2014)

15. Japan Aerospace Exploration Agency (JAXA). Product Format Description - LISM(TC,MI,SP)/SPICE, http://l2db.selene.darts.isas.jaxa.jp/help/en/LISM_SPICE_Fromat_en_V01-03.pdf

16. National Aeronautics and Space Administration (NASA). Lunar Mapping and Modeling Project (LMMP), http://pub.lmmp.nasa.gov/LMMPUI/LMMP_CLIENT/LMMP.html (accessed January 27, 2014)

17. National Aeronautics and Space Administration (NASA). The Structure of PDS Data Sets, http://pds.nasa.gov/data/data-structure.shtml (accessed January 27, 2014)

18. Open Geospatial Consortium (OGC). Web Map Service (WMS), http://www.opengeospatial.org/standards/wms (accessed January 27, 2014)

19. Oracle Corporation. Class HTMLEditorKit, http://docs.oracle.com/javase/7/docs/api/javax/swing/text/html/HTMLEditorKit.html (accessed January 27, 2014)

20. Oracle Corporation. JavaServer Pages Technology, http://www.oracle.com/technetwork/java/javaee/jsp/index.html (accessed January 27, 2014)

21. Tsunokake, T.: Moon Seeker: Search System for a Lunar Geographic Information System with a Query-By-Object Interface. Master's thesis, The University of Aizu, (March 2010), http://datadb04:8090/moon_seeker

22. The Apache Software Foundation. Apache Tomcat, http://tomcat.apache.org/. (accessed January 27, 2014)

23. The Eclipse Foundation. Eclipse, http://www.eclipse.org/ (accessed January 27, 2014)

24. The Open Source Geospatial Foundation (OSGeo). MapServer, http://mapserver.org/ (accessed January 27, 2014)
25. The OpenLayers Dev Team. OpenLayers, http://openlayers.org/ (accessed January 27, 2014)
26. The OpenLayers Dev Team. OpenLayers.Geometry.Polygon, http://dev.openlayers.org/docs/files/OpenLayers/Geometry/Polygon-js.html (accessed January 27, 2014)
27. The PostgreSQL Global Development Group. PostgreSQL, http://www.postgresql.org/ (accessed January 27, 2014)
28. The Seasar Foundation. S2JDBC, http://s2container.seasar.org/2.4/en/s2jdbc.html (accessed January 27, 2014)
29. The Seasar Foundation. Seasar2, http://www.seasar.org/en/ (accessed January 27, 2014)
30. Washington University in St. Louis. Lunar Orbital Data Explorer, http://ode.rsl.wustl.edu/moon/ (accessed January 27, 2014)
31. World Wide Web Consortium (W3C). CGI: Common Gateway Interface, http://www.w3.org/CGI/ (accessed January 27, 2014)
32. Yahoo! Inc. Yahoo! Maps, http://maps.yahoo.com/ (accessed January 27, 2014)
33. Yukihiro Matsumoto. Ruby, https://www.ruby-lang.org/ (accessed January 27, 2014)

Search System for City Information Using Multiple Public Transportation Information Resources

Wanming Chu

Aizu University
Aizu-Wakamatsu, 965-8580, Japan
w-chu@u-aizu.ac.jp

Abstract. In this paper, we propose a novel search system for city public transportation. This system collects the bus timetable information from Aizu bus home page, the city information from Internet, and the public facilities map of Aizu-Wakamatsu city, to generate geographic data by using Google Maps API. Multiple search methods are developed to obtain the information that users are interested in. One of the search results can be set as the origin or destination of a bus route. The shortest bus route with the minimum number of bus stops between the origin and destination can be found by using the bus routing function. The search results and the shortest bus route are visualized on the embed Google map(interactively). The search detailed information are shown in the side-bar. Since this system finds city information with bus route, residents and visitors can utilize the city public transportation more efficiently for their daily life, business, and travel planners.

1 Introduction

Nowadays, people can use Google map to find the shortest route for car, train, or walk with two input addresses or screen locating. Google map offers Ajax technology to show the route information details and route direction on the map in one page. By using an installed car navigation, drivers can easily find unfamiliar destinations that, they want to visit. For residents, bus service is the most important means of transportation because of its low cost, convenience of time schedule and reasonable bus line arrangement. Most cities distribute bus time schedule handbooks or open an online page to help residents or visitors to search bus time table and bus route. There are kinds of web-based bus route search system[1][2][3] like Google map route search method. These web-based search systems mainly contribute to find bus routes between two bus stops depending on fare, time, or distance. This search method is not efficient when residents or visitors need to go to the place where they have not visited yet. There is a need for extra search method to find the destination information using Google or some service call. For example, a patient wants to change from one hospital to another hospital[4], a visitor needs to visit city hall from hotel

A. Madaan, S. Kikuchi, and S. Bhalla (Eds.): DNIS 2014, LNCS 8381, pp. 256–265, 2014.

for business, and a traveler wants to find a hotel close to a supermarket from the station. Since existing online search systems do not have enough city information and search methods, residents or visitors can not efficiently utilize public transportation. Our proposal is to implement the public transportation search system that should be provided with the following functions.

- Users can use it to find bus line and bus stop information, and find the bus route between origin and destination bus stops/places like Google map search function.
- Users can find the bus route by navigating the source and destination facilities in the city like car navigation.
- Users can find the bus route with complicated query through a simple query interface.

The rest of this paper is organized as follows. In section 2, we describe the system architecture and search interface layout. In section 3, we explain the database design and geometry data types. In section 4 we introduce the bus routing algorithm. In section 5 we demonstrate our three search interfaces. We conclude and discuss the future work in Section 6.

2   Architecture and Layout

Figure 1 shows the system architecture. The system is implemented by PHP 5.3, PostgreSQL 9.0, jQuery1.8, HTML5, and Google Maps JavaScript API V3. The client-side provides geographic data generation and search interfaces that

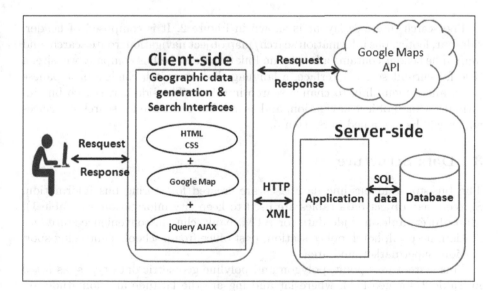

Fig. 1. System Architecture

use Google Map API and jQuery AJAX to get xml data from the server-side and present the data on the map by makers, polygon, polyline, and direction. The server-side uses PHP language to implement a web application. The PHP application generates SQL from input parameters and accesses database. The acquired data from database will be formed into xml format data and sent to the client-side.

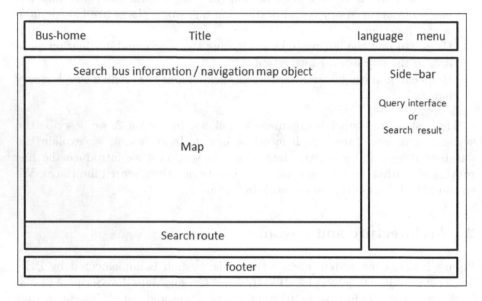

Fig. 2. Search Interface Layout

The search interface layout is shown in Figure 2. It is composed of header, side-bar, footer, bus information search/map object navigation, route search, and map. The header contains a bus-home link to access the bus company website, a title for current search interface, a language link to switch English or Japanese page, and a menu link to change search interface. The side-bar, search bus information/map object navigation, and search bus route are for search interfaces which will be explained in Section 5.

3   Data Structure

The bus-stop and bus-line databases are created for storing bus information. Sixteen map object data tables are used to keep city information(see Table 1). Map object tables include data for ATM, bank, church, convenience store, gas station, hospital, hotel, police station, post office, road, school, sightseeing spot, station, supermarket, and area.

This system uses point, polygon and polyline geometric data types, as listed in Table 2, for geo field, where lat and lng are the latitude and longitude, respectively, of a point on Google Map.

Table 1. Tables in Database

Category	Table	Fields
Bus information	bus-stop	id,name,address,geo
	bus-line	id,name,num,busstopId,departureTime
City information	Map object	id,name,[category],address,geo

Table 2. Geometric Data Types

Type of geo	Format	Map object
point	(lat,lng)	others
polygon	(lat1,lng1),(lat2,lng2),...)	area
polyline	(lat1,lng1),(lat2,lng2),...)	road

4   Routing

We adopt SQL to implement direct, one transfer, and two transfer route algorithms between A and B, where A is the origin bus stop and B is the destination bus stop. A.num is the sequence number of the origin bus stop in a bus line. B.num is the sequence number of the destination bus stop in a bus line. A transit bus stop passes multiple bus lines. The algorithms are shown below.

```
Direct(A,B){
   Find all buslines that pass A and B;
   Find one busline with minimum(B.num - A.num);
   Display  busline,  busstops from A to B;
 }
One_transfer(A,B){
   Find all transit busstops;
   Find all buslines that pass A and transit[i], where i=0-n;
   Find all buslines that pass transit[i] and B, where i=0-n;
   Find one busline with minimum(transit[i].num-A.num)+B.num-transit[i].num);
   Display busline1, bustops from A to transit[i];
   Display busline2, bustops from transit[i] to B;
}
Two_Transfer(A,B){
   Find all transit1 busstops from A;
   Find all transit2 busstops to B;
   Find all buslines that pass A and transit1[i], where i=0-n;
   Find all buslines that pass transit1[i] and transit2[j], where i=0-n, j=0-m;
   Find all buslines that pass transit2[j] and B, where j=0-m;
   Find one busline with minimum(transit1[i].num-A.num)+B.num-transit2[j].num);
   Display busline1, busstops from A to transit1[i];
   Display busline2, busstops from transit1[i] to transit2[j];
   Display busline3, busstops from transit2[j] to B;
}
```

5   Search Interfaces

There are three kinds of search interfaces in this system. The bus tracker interface provides search functions based on the bus information. The map object search interface navigates city information. The query-by-object interface selects two objects and one geographic operator to build complicated SQL for information search. All the search results can be set as an origin or a destination for bus route search. A top menu allows users to select one of them.

5.1   Bus Tracker Interface

There are two search functions and three bus route search methods in bus Tracker interface. A user can select one of the bus line from the bus line list. Bus line search function lists all bus stops on side-bar, visualizes the numbered markers for all bus stops and the direction of the bus line on the map. Bus stop search lists all bus lines that pass this stop on side-bar and a marker on map for this bus stop. Each marker on the map has a information window with two set buttons for setting this marker as departure bus stop or destination bus stop.

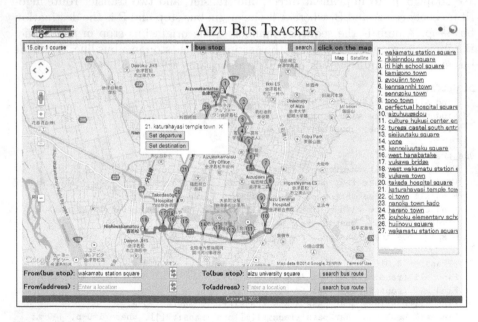

Fig. 3. Bus Tracker Search

Users can search bus route by inputting/setting from bus stop and to bus stop, by inputting two addresses, or by clicking any two point on the map. In Figure 3, City-1 course is selected and the 21st bus stop is set as the departure bus stop. In Figure 4, the shortest bus route from Katurahayasi Temple Town to Aizu University is shown on the map and the bus stop information of this bus route are listed as links on the side-bar.

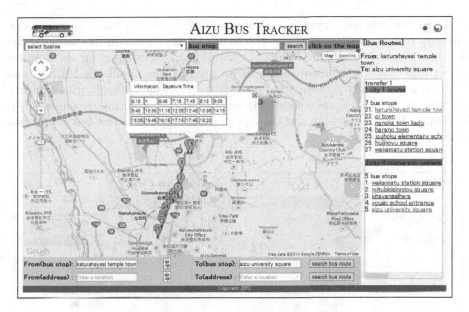

Fig. 4. Bus Tracker Routing

5.2   Map Object Search Interface

Bus route search by map object interface provides a navigation function to search city information. In this interface, a user first selects one object from the map object list.

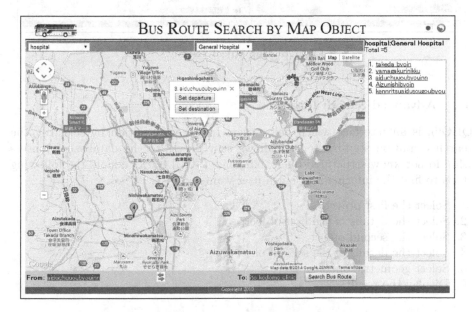

Fig. 5. Map object search

When the selected object has a category attribute, the user can select one category from a category list. The object details are shown on the map and the side-bar. Figure 5 shows the results of general hospital category for the hospital object and sets Aiduchuuoubyouinn as an origin place. Similarly, Ito Kodomo Clinic is selected as the destination place from the results of pediatrics category for hospital object. Figure 6 shows the bus route from Aiduchuoubyouinn to Ito Kodomo Clinic on the map and the bus stop information of this bus route with links on the side-bar.

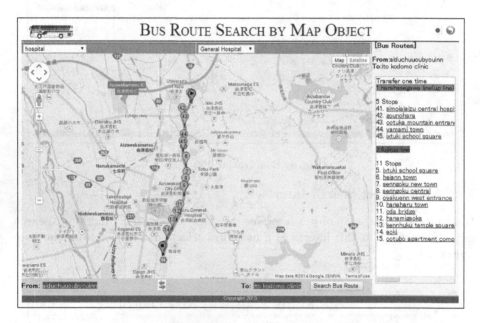

Fig. 6. Map object routing

5.3   Advanced Search Interface

QBO[5] is an interface which helps users to get information by selecting the search condition step-by-step. It allows easy access to the database for users who do not know structured query language (SQL). QBO adopts the following steps to find the information that a user wants.

1. Select the first object.
2. Select the details of the first object.
3. Select the second object.
4. Select the details of the second object.
5. Select geometric operators.
6. Display the result.

Table 3. Geometric Operators

Geometric operators	Geometric operators in PostgreSQL	Data type	Description
near	↔	Point @ Circle	Distance
contain	?#	Path ?# Path	Intersect
	~	Circle ~ Point	Contain
	~=	Point ~= Point	Same as
disjoin	not ?#	not (Path ?# Path)	not Intersect
	not ~	not (Circle ~ Point)	not Contain
	not ~=	not (Point ~= Point)	not Same as
within	@	Point @ Circle	Inside

Our system defines four geographic operators: near, contain, disjoint, and within. They are implemented by geometric operators in PostgreSQL, as listed in Table 3.

In order to search a bus route from Aizu University to a post office which is near a Liondor supermarket, we can use the QBO interface to find the destination. Figure 7 shows the first page of QBO in the side-bar. It displays all of the objects in the database. The post office object is selected as the first object.

The remaining five steps are shown in Figure 8. They are selecting all details of the post office object, selecting second object, selecting supermarket object, selecting Liondor from the details of the supermarket object, selecting near operation, and displaying the results. From one step to next step one submit button should be pushed.

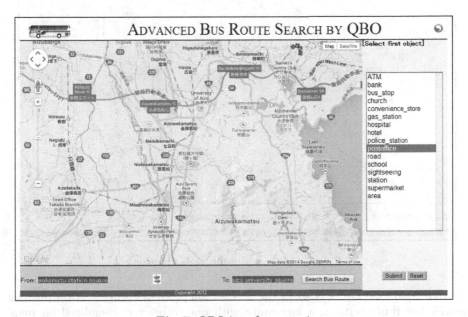

Fig. 7. QBO interface: step1

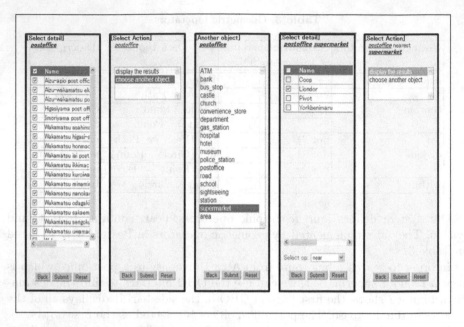

Fig. 8. QBO interface: step2 - step6

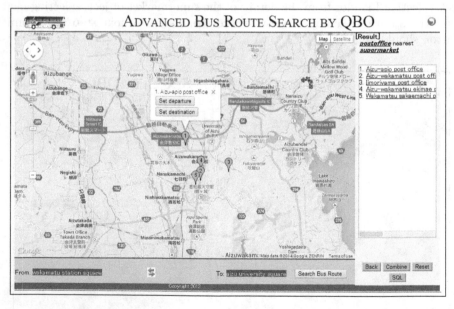

Fig. 9. Set search result as destination

Aizu-Apio Post Office is set as the destination place in Figure 9. After pushing search bus route button, a bus route with one-transfer is visualized on map, shown as in Figure 10.

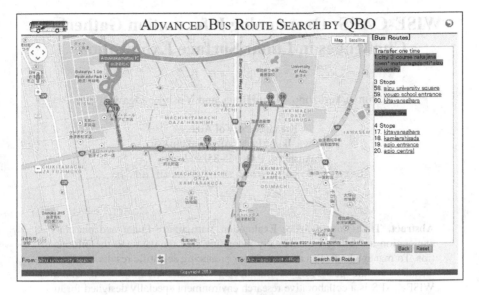

Fig. 10. QBO routing

6   Summary and Conclusions and Future Work

The study describes a public transportation search system that has been developed recently. With city information and multiple search methods, the system can be used for online city information guidance or special purposes, for example, as a referral hospital guidance. It also can be used as a aid tool for personal travel/business planers. Future study and research will focus on developing web service for mobile phone Android applications.

References

1. Hoar, R.: Visulizing transit throught a web based geographic information system. Internatianl Jounal of Human and Social Sciences 4(8), 607–612 (2009)
2. Kumar, P., Reddy, D., Singh, V.: Intelligent transportation system using gis. In: Proc. Map India Int. Conf. GIS, GPS, Aerial, Photography, and Remote Sensing (2003)
3. Pun-Cheng, L.S.C.: An interactive web-based public transport enquiry system with real-time optimal route computation. IEEE Transactions on Intelligent Transportation Systems 13(2), 983–988 (2011)
4. Kobayashi, S., Fujioka, T., Tanaka, Y.: A geographical information system using the google map api for guidance to referral hospitals. J. Medical Systems 34(6), 1157–1160 (2010)
5. Rahman, S.A., Subhash Bhalla, T.H.: Query-by-object interface for information requirement elicitation in m-commerce. Int. J. Hum. Comput. Interaction 20(2), 135–160 (2006)

WISE-CAPS: Overcoming Information Gathering Challenges in Lunar Surface Exploration

Junya Terazono, Naru Hirata, and Yoshiko Ogawa

The University of Aizu,
Tsuruga, Ikki-Machi, Aizu-wakamatsu,
Fukushima, P.O. 965-8580, Japan
{terazono,naru,yoshiko}@u-aizu.ac.jp

Abstract. Time-domain Space Exploration, particularly Lunar and planetary exploration are facing exponentially increasing size the data and the information. To manage huge data and to help for producing scientific results, we focus attention on the possibility of introducing GIS and creating WISE-CAPS. WISE-CAPS is a collaborative research environment specially designed for lunar and planetary exploration domain.

1   Introduction

The increase of data is eminent in Lunar and Planetary Exploration field. In 1994, NASA (National Aeronautics and Space Agency) and DoD (Department of Defense) launched Clementine [1], small-weight lunar explorer and made orbital survey for two months. This spacecraft is the first in the world which acquired full lunar surface photographs in digital format. All raw data were stored in 88 CD-ROMs and therefore the full data amount was approximately less than 60 Gigabytes. In September 2007, the first Japanese large-scaled lunar explorer, called "Kaguya" (also as SELENE, SELenological and Engineering Explorer) launched. Kaguya orbited around the moon for approximately 20 months and made explicitly detailed observations of the moon. The amount of data returned from the moon is approximately 15 Terabytes, and processed data called "Level 2" is approximately 50 TB [2]. Lunar Reconnaissance Orbiter, the newest orbiter which has been announced in September 2013, sends image data acquired and stored in the PDS, Planetary Data System, it has reached 133 terabytes [3] of size. Thus, the amount of data being handled in lunar exploration is rapidly increasing from gigabyte order to terabytes order in these twenty years. The users face petabyte data in near future as data is cumulating as mission is flawlessly operating.

This rapid increase of data is partly due to advancement of sensor technology, and partly due to improved efficiency and wider frequency availability of transmission from celestial bodies to the Earth. As sensor technology is advancing, it is expected that increasing amount of data will be sent from the moon and for other planetary exploration.

A. Madaan, S. Kikuchi, and S. Bhalla (Eds.): DNIS 2014, LNCS 8381, pp. 266–273, 2014.
© Springer International Publishing Switzerland 2014

1.1   Adverse Impact of Information Explosion in Science

Huge amount of data in science inevitably brings the following problems in scientific field:

1. The data becomes too much for individual scientists. For example, we cannot check all image data acquired by Kaguya by one person, as it is expected to take more time compared to average life time of humans. It also means analysis and processing of huge data are beyond human resource. Of course, these processes can be automated in some extent, the huge data consumes time for transfer and handling. It is common, that - scientists first download data from data distribution sites, and then analyze them on their desktop. However, they spend more and more time in downloading (data transfer).
2. Versioning problem. Scientists produces several different versions for applying several different parameters for same original data. For example, ratio image, false color image by dividing a image of a wavelength with another and applying RGB color for three different dividing colors, can be created with different combination of selected wavelength (or "band") from the same image data set. This fact means plural data sets can exists from original data sets. Of course, this number will in-crease by further image data processing. If number and amount of data in-creases, version and variation of data may exceed beyond human ability to manage.
3. Adaptability for mobile devices. Recently, mobile device such as smart-phones and tablets have been introduced in research. These gadgets are generally used mostly as tools (reading/writing E-mails, voice communications, text transactions), however, can be used as a replacement of desktop computer even in re-search field in near future considering their rich computing power. These devices, however, lacks sufficient local storage to stock scientific data. Therefore, scientists should use these tools with high speed network for research, particularly frequent data uploading and download-ing. It is very unlikely to store and analyze scientific data mobile devices despite in-creasing speed of wireless network and CPU capability if amount of data increases more and more in the future.   Of course, research activity can be made by mainly per-sonal computers (regardless of their portability), however, it would be more useful if mobile device can be used not only for communication use but research itself.

We need to consider countermeasures to adapt more amount of scientific data in near future with currently existing ones reckoning above mentioned points in scientif-ic research.

2   The Example of Web-GIS Based Challenge: WISE-CAPS

2.1   Outline of WISE-CAPS

WISE-CAPS (Web-based Secure Environment for Collaborative Analysis of Planetary Science) [5, 6,7] is a Web-GIS based platform for data sharing, analysis and communi-cation environment specially designed for lunar and planetary exploration data. We explain our study about WISE-CAPS concept, structure and examples. WISE-CAPS aims to capture for all research-related activities to be completed inside the network, especially web browsers. Therefore, the system should have collaboration function to share, browse and limit the data. Also the system will have a communication function to enable on-line discussion based on data in WISE-CAPS. As WISE-CAPS postulates user-friendly design, users do not need to master particular programming languages and macros to use it. All WISE-CAPS function is realized using native function of web browsers (JavaScript, server side programming, and other web technology) and this concept avoids involuntary system instability caused by web browser plug-ins.

WISE-CAPS hardware consists free software on two servers, one for web and one for data-base. Apache httpd and PostgreSQL are used as basic server software respectively. We use MapServer [5], map data synthesizer for web, and OpenLayers [6], display framework for Ajax-compliant web page manipulator written by JavaScript, as middle software. All data transaction is compliant with OGC (Open Geospatial Consortium) [7] protocol. WMS (Web Mapping Service) [8] and WFS (Web Feature Service) [9] is basically used in WISE-CAPS and these open features enables smooth exchange with other server which can understand WMS/WFS protocols.

WISE-CAPS has its own user control mechanism. It is based on GridSite [10], installed as Apache httpd module, and uses GACL (Grid Access Control List). The GACL determines whether a user can access to the specific directory based on user's digital certificate. Once user installs his or her digital certificate, user can make use of user control function equipped with WISE-CAPS on every browser, even it is installed in mobile devices. However, we are now planning to replace this user control function from this GridSite-based system to Single Sign-On mechanism, as GridSite module is compatible only with Apache httpd 2.0 series which is currently out of support.

2.2   Current Achievement

Figure 1 shows typical screenshot of WISE-CAPS displaying overlap of two maps. In Figure 1, a deeper image region shows data obtained by Kaguya, more precise (higher resolution) than the base map which is based on images acquired by Clementine mission. Opacity of image can be controlled by controlling function in the web page (not shown below).

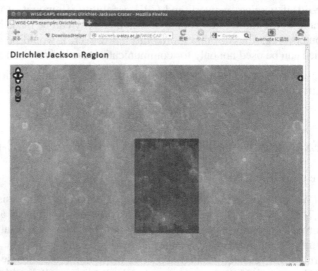

Fig. 1. WISE-CAPS mapping example of lunar Dirichlet and Jackson craters. Detailed (grayer) image shows recent and more precise image acquired by Kaguya mission. Base map uses UVVIS (UV-Visible) images obtained by older Clementine mission. Figure after [4].

Figure 2 shows more complex mapping. The figure shows mapping in WISE-CAPS in the Jack-son crater area on the moon. There are four layers in this figure, base map, TC (Terrain Camera) image obtained by Kaguya, MI (Multiband Imager) image also obtained by Kaguya, and a ratio image from MI data (color). We see only three parts in this figure MI panchromatic image and ratio image are totally overlapping. Each layer can set individual opacity settings using same controlling system stated in Figure 1. Opacity control is shown at the bottom of lunar image in this figure. Thanks to user controlling mechanism implemented on WISE-CAPS, MI panchromatic image and ratio image can be shown by authorized persons to show.

By overlapping several kind of image such as spectral data and topographic data, researchers can know and find relationships of spectra (it is strongly related to mineral and element composition of the area) and topographic features.

Third example is relatively complex. This is a integrated display system of lunar imagers. There are three imaging-related instruments on Kaguya, MI, SP (Spectral Profiler) and TC. These three instruments are called as LISM (Lunar Imager and Spectrometer) as a generic term. TC and MI captured raster data as shown in previous figures, and SP obtained high resolution spectral data (in wavelength) in one spot. Scientists working on Kaguya data awaits for integrated display system of these three instruments to know overall information of lunar surface science. However, it was difficult due to difference of nature of data. Here in Figure 3 shows integrated LISM data display system created by full function of WISE-CAPS.

There are two screens in Figure 3. Left screen shows small portion of the lunar surface. Here left screen shows MI ratio image and SP footprint and spots of spectrum measurement, as well as base layer (hidden in this figure). By clicking SP measurement spot, the spectral data are shown in the right graph area. Data of different spots are shown in different colors. In this system, raster data (MI and TC) are stored within

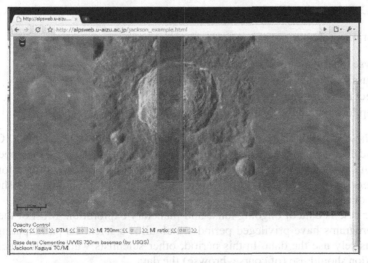

Fig. 2. Another WISE-CAPS display snapshot of Jackson crater region on the moon. There are four layers in this screenshot, base map (same as ones used in Figure 1), TC images (deeper gray image in this web screen), MI panchromatic image and MI ratio image (color). Figure after [4].

web server of WISE-CAPS and delivered using WMS proto-col as same as previous examples. Meanwhile, all SP data shown here are stored in database (PostgreSQL and PostGIS) and brought to web server using WFS protocol.

MapServer creates raster image with overlapped layers showing SP footprint for this web page. Plot of reflectance (SP data) are created using DyGraphs [11], Java-Script-based data visualization and graph creation library.

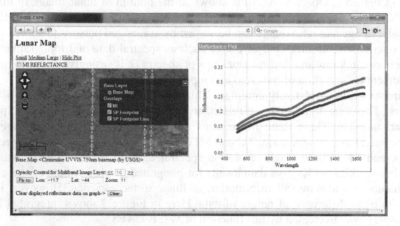

Fig. 3. Integrated data display system based on implementation of WISE-CAPS function. Data shows image and spectral data around Tycho crater on the moon. Figure after [4].

3 The Web-GIS Based Challenge: WISE-CAPS

3.1 Collaborative Platform

Collaboration refers to:

— Browse other persons' data (if permitted).
— Discuss with other persons using in this system.
— Protect his or her data from unauthorized access (browsing).
— Group forming and working.

All these capabilities require some personal identification mechanism. Generally, scientific data is open to anyone or public. However, in some cases, they need to keep these data as secret (or protected) position. For example, processed data for publishing papers should be hidden before publication. Such data should be only sharable for a group of authors.

Other case is data of ongoing lunar and planetary exploration. Most of these exploration programs have privileged period which allows scientists participating in them to exclusively use the data. In this period, other scientists who do not touch into the exploration should use (of course, browse) the data.

However, it often requires in scientists' group(s) who involves in the same exploration program to share the result of data for examination of further exploration. Without definite authentication mechanism, system cannot protect data and people using system and data. Also it prevents "data over-sharing". Oversharing means that data which should be kept inside the small group is spilled over the public or external area. Such oversharing is generally due to mis-setting of sharing range and sharing operation.

Therefore, WISE-CAPS should have the following mechanism to protect data.

-      Mechanism that is able to identify each individual.
-      Mechanism that can for multiple groups.
-      Mechanism that is used for several servers.
-      Mechanism that is valid for multiple applications.
-      Mechanism that is easy-to-use for both administrators and end users.

WISE-CAPS system used to utilize GridSite [12] module for user authentication mechanism mostly used for computational grid. GridSite is used as a loadable module of Apache httpd [13] and understand GACL (GridSite Access Control Language), XML based access control language, to assess whether a user can access under a directory. User identification is made by digital certificates, and password authentication is also available via the proxy server.

However, this GridSite module is compatible for only Apache httpd 2.0 series which is now out of support. Moreover, as this system is module basis, installation of multiple servers are required, and users should authenticate in each server. We are now considering another and better authentication mechanism to rise above its weakness to adopt OpenAM [14], open source authentication mechanism.

OpenAM, formerly named as OpenSSO, is a platform to enable simple and robust authentication and access control mechanism. As single-sign-on mechanism is realized, users can use only one authentication to use all resources spotted among WISE-CAPS system. Also, OpenAM interactively works with LDAP (Lightweight Directory Access Protocol) which is becoming de facto standard for resource management such as user ID, password and mail ad-dress, and therefore resource administration is relatively easy compared to GridSite, which re-quires independent user control mechanism with the operating system, once user administration is unified into LDAP.

Thus, OpenAM is friendly for both end users and administrators. We are now starting implementation of this new user control mechanism to WISE-CAPS and making demonstration system to evaluate its superiority.

3.2   Key Technology to Enable Our Demands

It would be convenient if the system has a mechanism to interact with others. For example, other system can display layers hosted in our WISE-CAPS, and users can include small maps using our data in their web pages and blogs as many users do using Google Map or other location-based web service. Already the system can interact with other system as WISE-CAPS uses open standard compliant. WMS, WFS protocol and OpenLayers can help display our layers to web pages in other system. Reversely, WISE-CAPS has a capability to display layers (or data) in other system via WMS/WFS protocol and OpenLayers. However, we need more functions to provide easier interaction with web pages.

This can be realized by API, Application Program Interface. API is becoming common in web service such as Google and Yahoo!, and uses small pieces of Java-Script to bring its function to their web pages. Users can use many functions by writing small JavaScript code inside the web page using APIs

— Users do not need to know internal mechanism, function and protocols of WISECAPS. What user only have to do is to know API name, parameter and format to acquire any required information from WISE-CAPS.
— Users can only use JavaScript to retrieve data from WISE-CAPS. No other software or programming are not required.
— For administrators of WISE-CAPS, they do not need to know inside architecture to public or at least those who want to use its function. It is also useful for them to hide changes of internal architecture.

Currently, prototype API is implemented. This API just shows arbitrary part of lunar image base map of WISE-CAPS in web browsers. This function is nearly same as one used in Google Map API to display its map on web browser. Currently, the designing of API structure and format is in progress. Also, we are investigating what function is required from researchers as API implementation.

4   Summary and Conclusion

WISE-CAPS, as an example of a solution for huge amount of scientific data for researchers. Location based data arrangement and web based platform (web browser as an entrance for all activity of research) is current our answer to overcome the increasing data in lunar and planetary exploration region. WISE-CAPS is a test bed for implementation of technology for adapting to large amount of data. We anticipate that the technologies demonstrated in here will be implemented in wider platform worldwide and connect them via open protocol and network to enable planetary scientists ease to wield large amount of exploration data.

References

1. Nozette, J., Rustan, P., Pleasance, L.P., Kordas, J.F., Lewis, I.T., Park, H.S., Priest, R.E., Horan, D.M., Regeon, P., Lichtenberg, C.L., Shoemaker, E.M., Eliason, E.M., McEwen, A.S., Robinson, A.S., Spudis, P.D., Acton, C.H., Buratti, B.J., Duxbury, T.C., Baker, D.N., Jakosky, B.M., Blamont, J.E., Corson, M.P., Resnick, J.H., Rollins, M.E., Davies, M.E., Lucey, P.G., Malaret, E., Massie, M.A., Pieters, C.M., Reisse, R.A., Simpson, R.A., Smith, D.E., Sorenson, T.C., VorderBreugge, R.W., Zuber, M.T.: The Clementine Mission to the Moon: Scientific Overview. Science 266(5192), 1835–1839 (1994)
2. Hoshino, H., Yamamoto, Y., Sobue, S., Yonekura, K., Ogawa, M., Iwana, Y., Matsui, K., Okumura, H., Kato, M.: Data Processing at KAGUYA Operation and Analysis Center. Space Science Reviews 154(1-4), 317–342 (2010)
3. LROC 15th PDS release, Lunar Reconnaissance Orbiter Camera news, http://lroc.sese.asu.edu/news/index.php?/archives/810-LROC-15th-PDS-Release.html

4. Terazono, J., Nakamura, R., Kodama, S., Yamamoto, N., Hirata, N., Ogawa, Y., and De-
    mura, H.: WISE-CAPS: Browsing and Analyzing System for Lunar and Planetary Data.
    Journal of Space Science Informatics Japan (2), 89–102, 2013. (in Japanese with English
    abstract)
5. MapServer, http://www.mapserver.org/
6. OpenLayers, http://www.openlayers.org/
7. OGC, http://www.opengeospatial.org/
8. Web Mapping Service, http://www.opengeospatial.org/standards/wms
9. Web Feature Service, http://www.opengeospatial.org/standards/wfs
10. GridSite, http://www.gridsite.org/
11. DyGraphs, http://www.dygraphs.com/
12. GridSite, http://www.gridsite.org/
13. Apache httpd, http://httpd.apache.org/
14. OpenAM, http://forgerock.com/products/
    open-identity-stack/openam/

Author Index